DATE DUE

OCT 17 2013			
GAYLORD			PRINTED IN U.S.A.

Biological Monitoring:
Theory & Applications

WIT*PRESS*

WIT Press publishes leading books in Science and Technology.
Visit our website for the current list of titles.
www.witpress.com

WIT*eLibrary*

Home of the Transactions of the Wessex Institute, the WIT electronic-library provides the
international scientific community with immediate and permanent access to individual
papers presented at WIT conferences. Visit the WIT eLibrary at
http://library.witpress.com

The Sustainable World

Aims and Objectives

Sustainability is a key concept of 21st century planning in that it broadly determines the ability of the current generation to use resources and live a lifestyle without compromising the ability of future generations to do the same. Sustainability affects our environment, economics, security, resources, health, economics, transport and information decisions strategy. It also encompasses decision making, from the highest administrative office, to the basic community level. It is planned that this Book Series will cover many of these aspects across a range of topical fields for the greater appreciation and understanding of all those involved in researching or implementing sustainability projects in their field of work.

Topics

Data Analysis
Data Mining Methodologies
Risk Management
Brownfield Development
Landscaping and Visual Impact Studies
Public Health Issues
Environmental and Urban Monitoring
Waste Management
Energy Use and Conservation
Institutional, Legal and Economic Issues
Education
Visual Impact

Simulation Systems
Forecasting
Infrastructure and Maintenance
Mobility and Accessibility
Strategy and Development Studies
Environment Pollution and Control
Land Use
Transport, Traffic and Integration
City, Urban and Industrial Planning
The Community and Urban Living
Public Safety and Security
Global Trends

Main Editor

E. Tiezzi
University of Siena
Italy

D. Almorza
University of Cadiz
Spain

M. Andretta
Montecatini
Italy

A. Bejan
Duke University
USA

A. Bogen
Down to Earth
USA

I. Cruzado
University of Peurto Rico-Mayazuez
Puerto Rico

W. Czyczula
Krakow University of Technology
Poland

M. Davis
Temple University
USA

K. Dorow
Pacific Northwest National Laboratory
USA

C. Dowlen
South Bank University
UK

D. Emmanouloudis
Technical Educational Institute of
Kavala
Greece

J.W. Everett
Rowan University
USA

R.J. Fuchs
United Nations
Chile

F. Gomez
Universidad Politecnica de Valencia
Spain

K.G. Goulias
Pennsylvania State University
USA

A.H. Hendrickx
Free University of Brussels
Belgium

I. Hideaki
Nagoya University
Japan

S.E. Jorgensen
The University of Pharmeceutical
Science
Denmark

D. Kaliampakos
National Technical University of
Athens
Greece

H. Kawashima
The University of Tokyo
Japan

B.A. Kazimee
Washington State University
USA

D. Kirkland
Nicholas Grimshaw & Partners
UK

A. Lebedev
Moscow State University
Russia

D. Lewis
Mississippi State University
USA

N. Marchettini
University of Siena
Italy

J.F. Martin-Duque
Universidad Complutense
Spain

M.B. Neace
Mercer University
USA

R. Olsen
Camp Dresser & McKee Inc.
USA

M.S. Palo
The Finnish Forestry Research
Institute
Finland

J. Park
Seoul University
Korea

M.F. Platzer
Naval Postgraduate School
USA

V. Popov
Wessex Institute of Technology
UK

A.D. Rey
McGill University
Canada

H. Sozer
Illinois Institute of Technology
USA

A. Teodosio
Pontificia Univ. Catolica de Minas
Gerais
Brazil

W. Timmermans
Green World Research
The Netherlands

R. van Duin
Delft University of Technology
The Netherlands

G. Walters
University of Exeter
UK

Biological Monitoring:
Theory & Applications

BIOINDICATORS AND BIOMARKERS FOR ENVIRONMENTAL
QUALITY AND HUMAN EXPOSURE ASSESSMENT

EDITOR

M.E. Conti
University di Roma 'La Sapienza', Italy

WIT*PRESS* Southampton, Boston

Editor

M.E. Conti

Universita di Roma 'La Sapienza', Italy

Published by

WIT Press

Ashurst Lodge, Ashurst, Southampton, SO40 7AA, UK
Tel: 44 (0) 238 029 3223; Fax: 44 (0) 238 029 2853
E-Mail: witpress@witpress.com
http://www.witpress.com

For USA, Canada and Mexico

WIT Press

25 Bridge Street, Billerica, MA 01821, USA
Tel: 978 667 5841; Fax: 978 667 7582
E-Mail: infousa@witpress.com
http://www.witpress.com

British Library Cataloguing-in-Publication Data

A Catalogue record for this book is available
from the British Library

ISBN: 978-1-84564-002-6
ISSN: 1476-9581

Library of Congress Catalog Card Number: 2007936323

*The texts of the papers in this volume were set
individually by the authors or under their supervision.*

Contents

Preface

Author Biographies

Acronyms

Preface

Environmental pollution, and the related human health concerns, has now reached critical level in several areas in the world. International programmes for researching, monitoring and preventing the causes of this phenomenon are ongoing in many countries.

There is, in fact, an imperative call for reliable and cost-effective information on the basal pollution levels of areas involved in intense industrial activities or sites of interest for possible future industrial development.

Within this framework, biomonitoring methods can be used as unfailing tools for the control of contaminated areas as well in environmental prevention studies (i.e. Environmental Impact Assessment or Strategic Environmental Assessment procedures). Human biomonitoring, in its turn, is now widely recognized as a mean tool for human exposure assessment, providing suitable and useful indications of the 'internal dose' of chemical agents. Bioindicators, biomonitors, and biomarkers are all terms well-known among environmental scientists, although their meaning is sometimes misrepresented. In this context, a better and a full comprehension of the role of biological monitoring and of its procedures in evaluating polluting impacts on environment and health is needed.

This book aims to give an overview of the current knowledge on relevant aspects of biological monitoring regarding the evaluation of ecosystems quality and human health.

In the first chapter the theoretical aspects of biomonitoring are presented. Types of biomonitoring, reference levels, concentration factors and the biomagnification phenomenon are fully discussed. Then, a chapter dealing with biomarkers in environmental monitoring is presented. It carefully analyzes the various kinds of potential biomarkers such as DNA or mRNA alterations, protein responses, metabolism products, hysto-cytopathological, immunological, physiological and behavioural biomarkers. The subsequent three chapters report the relevant biomonitoring methods and biomonitors that could be used in these studies. Chapter 3 deals with freshwater environments in the context of the Water Framework Directive(2000/60 EC), and reports the biomonitoring methods for water classification such as macrobenthic communities, water fauna and flora, and those based on the whole river ecosystem. Chapter 4, authored in concert with M. Iacobucci, analyzes the use of relevant marine organisms as biomonitors of heavy metal pollution, while Chapter 5 is dedicated to lichens as bioindicators of air pollution. Lichen bioindication methods such as the Index of Atmospheric Purity and transplants, and their accumulation capabilities of contaminants (i.e. heavy metals, sulphur, nitrogen and phosphorous compounds, ozone, fluorides and chlorides, radionuclides, etc.) are fully discussed.

Chapter 6, authored by A. Alimonti and D. Mattei, reports the state-of-the-art in the identification and use of biomarkers in human health. Human biomarkers are discussed on the basis of the category of the xenobiotic-source as well as the degree of information they can give (exposure, effect, or susceptibility).

Finally, the last chapter, authored in concert with M. Mecozzi, shows the capability of statistical multivariate approaches to elucidate and solve the problems related to the interpretation of experimental data obtained in biomonitoring studies. Practical and useful examples of these statistical capabilities are given.

I wish to thank the Authors for participating in this project, and for their important suggestions and comments. I also gratefully acknowledge Sergio Giovannini and Fabrizio Cecchini for their technical contribution. Finally, I wish to thank the Editors of Research Trends, and Elsevier, and Professor Dorgham of Inderscience Ltd, who kindly gave their permissions for the realization of this project.

Marcelo Enrique Conti

Author Biographies

Marcelo Enrique Conti is Professor of Environmental Impact Assessment at the University of Rome 'La Sapienza' and the University of Urbino, Italy. He has a doctorate in Food Control and Technology from the University of Rome 'La Sapienza'. His research interests encompass environmental fields including biological monitoring, analytical methods of quality control of the ecosystems and managerial problems concerning environmental heritage. He has held numerous lectures and didactic seminars, by invitation, in various Italian and overseas universities. He is a founding member and Vice-President of the International Academy of Environmental Sciences (IAES) at Venice. He is Editor-in-Chief of the International Journal of Environment and Health (IJENVH) of Inderscience Publishing Ltd and component advisor of the Scientific Committee of various international scientific journals. He is also author of more than 120 scientific publications including books, chapters of books and articles on national and international journals in the environmental field.

Mauro Mecozzi is Senior Researcher scientist at the Central Institute for Marine Research in Rome. His main fields of activity are the development of environmental analytical methods in spectroscopy and chromatography with related application to the monitoring of the marine environment and the application of univariate and multivariate statistical methods for the elaboration of environmental data. He is author of more than 60 papers in the field of environmental studies.

Alessandro Alimonti is Senior Researcher of the Department of Environment and Primary Prevention, Italian National Institute of Health. His research interests mainly focus on the role of chemical elements in health and environmental problems, encompassing the evaluation of markers of exposure, effect and susceptibility, the development of validated analytical methodologies, and the assessment of reference values for elements in biological materials for use in life sciences. He was/is a participant in and/or responsible for several National and European research projects. He is also the author or co-author of about 170 scientific contributions.

Daniela Mattei is a graduate, *cum laude*, in Biology from the University of Rome 'Tor Vergata'. She obtained her Bioinformatics Master degree, *cum laude*, from the University of Rome 'La Sapienza'. Currently she is pursuing her PhD in Chemical Sciences, by working with Dr Alimonti as Supervisor in the laboratories of the Bioelements and Health Unit of the Department of Environment and Primary Prevention, Italian National Institute of Health. Her PhD research is focused on the development of monitoring methods to evaluate the occupational exposure and the health risk by measuring and assessing chemical elements in tissues, secreta and excreta by ICP-MS and electrochemical biosensors.

Marta Iacobucci is a graduate in the Faculty of Environmental Sciences at the University of Urbino. Her research interests are in the area of biological monitoring. She held a seminar on biomonitors and their utility at the University of Rome in 2004. Currently she is pursuing her PhD in Chemical Sciences, by working with Professor Conti as Supervisor in the Center for the Environmental Assessment of Industrial Activities of the Faculty of Environmental Sciences at the University of Urbino.

Acronyms

ACGIH	American Conference of Governmental Industrial Hygienists
AchE	Acetylcholinesterase
AGD	Amoebic gill disease
Ah	Aryl Hydrocarbon (receptor)
ALAD	δ-aminolevulinic acid dehydratase
AOAC	Association of Official Analytical Chemists
AQCS	Analytical Quality Control Service
ATP	Adenosine Tri-Phosphate
BAT	Biologische Arbeitsstofftoleranzwerte, i.e., DFG Biological Tolerance Values
B[a]P	benzo[a]pyrene
BchE	butyrylcholinesterase
BDI	standardized Biological Diatom Index – BDI (France)
BEI	Biological Exposure Index
B-Hg	Mercury concentration in blood
BL	Background Level
BNF	β-naphthoflavone
BOD$_5$	a five day Biological oxygen demand
B-Pb	Lead concentration in blood
CAT	Catalases
CbE	carboxylesterase
CBeD	Chronic Berillium Disease
CB	carbamates
CCA	Canonical Correspondence Analysis
CEE index	diatom Index of European Economic Community
CF	Concentration Factor
CF-IRMS	Continuous Flow-Isotope Ratio Mass Spectrometry
ChE	organophosphorous-sensitive cholinesterases
Chl	Chlorophyll
COD	Chemical Oxygen Demand
CRM	Certified Reference Material
CV	Coefficient of Variation
CYP2E1	Cytochrome P450 2E1

DAMD	Discriminant analysis by Mahalanobis distance
DBT	Dibutyltin
DDD	Dichlorodiphenyldichloroethane
DDE	Dichlorodiphenyldichloroethylene
DDT	Dichlorodiphenyltrichloroethane
DFG	Deutsche Forschungsgemeinschaft, i.e., German Research Foundation
DMSO	Dimethyl sulfoxide extraction technique
DO	Dissolved oxygen
DPSIR	Drivers-Pressures-State-Impact-Responses
DT-d	DT-diaphorase
DU	Depleted uranium
DW	dry weight
EBI	Extended Biotic Index
EEA	European Environmental Agency
EIA	Environmental Impact Assessment
ELISA	Enzyme-linked immunosorbent assay
EPA	Environmental Protection Agency
EPI-D	Eutrophication Pollution Index using Diatoms (Italy)
EPR	Electron Paramagnetic Resonance
EROD	7-ethoxyresorufin O-deethylase
ETAAS	Electrothermal Atomic Absorption Spectrometry
ETU	Ethylene Thiourea
FW	fresh weight
G6PHD	glucose-6-phosphate dehydrogenase
GC	Gas Chromatography
GDI	Generic Diatom Index (France)
GIS	Groupement d'Intéret Scientifique indexes
GPx	Glutathione peroxidase
GR	Glutathione reductase
GSH	Glutathione
GST	Glutathione S-transferase
HAC	Hierarchical Ascendant Classification
Hb-A	Haemoglobin Adducts
HCB	Hexachlorobenzene
HHC	Halogenated Hydrocarbons
HLA	Human Leukocyte Antigen
HLA-DP	protein/peptide-antigen receptor composed of 2 subunits, DPα and DPβ, encoded by two *loci*, HLA-DPA1 and HLA-DPB1 that are found in the Human Leukocyte Antigen complex on human chromosome 6
HPCD	Hydroperoxy conjugated dienes

HSP	Heat shock proteins
I.I.	Ichthyological Index
IAEA	International Atomic Energy Agency
IAP	Index of Atmospheric Purity
IARC	International Agency for Research on Cancer
IBE	Extended Biotic Index
IBI	Index of Biotic Integrity
IBL	Lichen Biodiversity Index (Italy)
IBMR	Biologique Macrophytique en Rivière index
ICI	Index of the Community of Invertebrates
ICOH	International Commission on Occupational Health
ICP-AES	Inductively Coupled Plasma - Atomic Emission Spectroscopy
IFF	Fluvial Functionality Index
ILM	Leclercq and Maquet Index (Belgium)
IP	index of poleotolerance
ISO	International Organization for Standardization
IUPAC	International Union of Pure and Applied Chemistry
LC	Liquid Chromatography
LC 50	Lethal Concentration (LC values usually refer to the concentration of a chemical in air but in environmental studies it can also mean the concentration of a chemical in water. For inhalation experiments, the concentration of the chemical in air that kills 50% of the test animals in a given time (usually 4 h) is the LC 50 value.)
LDA	Linear Discriminant Analysis
LDH	Lactated dehydrogenase
LMS	Lysosomal membrane stability
MAE	Microwave Assisted Extraction
MAK	Arbeitsplatz-Konzentration, i.e., Germany Maximum Allowable Concentration
MATLAB	MATrix Laboratory (high-level technical computing language)
MDA	Malondialdehyde
MeHg	Methylmercury
MFO	Mixed Function Oxygenase
MIS	Macrophyte Index Scheme
MM	Methyl Mercury
MPO	Myeloperoxidase
MTR	Mean Trophic Rank
MT	Metallothioneins
MW	Microwave-assisted acid digestion
NAA	Neutron Activation Analysis

NADPH	Nicotinamide adenine dinucleotide phosphate-oxidase
NAG	N-acetyl-β-D-glucosaminidase
NAT2	N-Acetyltransferase 2
NHANES	National Health and Nutrition Examination Survey (USA)
NIR	Near InfraRed
NMR	Nuclear Magnetic Resonance
NORMSEP	SEParation of the NORMally distributed components of size-frequency samples
NQO1	Quinone oxidoreductase
NRR	Neutral Red Retention
OBT	Organically bound tritium
OD	optical density
OECD	Organisation for Economic Co-operation and Development
OPs	organophosphorous
OSHA	Occupations Safety and Health Standards
PAHs	Polycyclic Aromatic Hydrocarbons
PCA	Principal Component Analysis
PCBs	Polychlorinated Biphenyls
PCDDs	Polychlorinated dibenzodioxins
PCDFs	Polychlorinated dibenzofurans
PEL	Permissible Exposure Limits (by OSHA)
Ph	Phaeophytin
PI	Pollution Index
PIXE	Proton Induced X-ray Emission
PMIs	Preliminary Multimetric Indices
PON1	Paraoxonae 1
PNP	p-Nitrophenol
POPs	Persistent Organic Pollutants
PROTEIN HC	α_1-microglobulin
PTs	Proficiency tests
QC	Quality Classes
RCE	Riparian Channel Environmental
RDA	Redundancy analysis
REACH	Registration, Evaluation, Authorisation of Chemicals
REEs	Rare earth elements
RM	Reference Materials
ROT	Saprobic index of Rott (Austria)
RPLI	Relative penis length index
RPSI	Relative Penis Size Index
SCE	Sister Chromatide Exchange
SE	Standard Error

SFG	Scope for Growth
SF-ICP-MS	Sector Field Inductively Coupled Plasma Mass Spectrometry
SHE index (Germany)	
SLA	Sládeček index (Czech Republic)
SOD	superoxide dismutase
SPI	Specific Polluosensitivity Index (France)
S-PMA	S-Phenylmercapturic Acid
SPME	Solid Phase Microextraction
SQI	Site Quality Index
SU	Systematic Units
TBT	tributyltin
TDI	Trend Detection Index
TDI	Trophic Diatom Index (United Kingdom)
TIM	Trophic index macrophyten
TLV-TWA	Threshold Limit Value - Time Weighed Average
TTC	Trophic Transfer Coefficient
TU	Toxic Unit approach
TWA	Time-Weighted Average
U-As	Arsenic concentration in urine
U-Hg	Mercury concentration in urine
UNEP	United Nations Environment Programme
VDSI	Vas deferens sequence index
WFD	Water Framework Directive
WHO	World Health Organization

1 Environmental biological monitoring

M E Conti

1 Introduction

The problem of environmental quality control is strictly related to the implementation of adequate experimental methods for the assessment of the actual state of health of ecosystems.

Ecotoxicology, as opposed to traditional toxicology, relates to a *series* of effects and responses to the contaminants of the ecosystem under study. Therefore, interpretation of the experimental data concerning the presence of contaminants in a given ecosystem requires data processing that is adequate to complex systems.

By *effect* we mean the kind of reaction shown by the ecosystem (not necessarily of a toxic nature); by *response* we mean the quantification (e.g. the percentage) of the kind of effect. The effects on the organisms may concern the reproductive system, motility, growth rate, etc.

Ecosystems, that consist of a biotic and an abiotic compartment, may respond in very different ways, and the onset of toxicity phenomena, as is known, may not occur until after some time.

However, we can assume that very low doses of contaminants do not generate negative effects on the organisms (biotic component). Every contaminant has a threshold, though, beyond which detoxification phenomena occur: these are just defense mechanisms that organisms may develop in the presence of these contaminants. Such mechanisms are often noticed in the different branches of the phylogenetic tree of the various species. In this respect, one of the most studied mechanisms is the one concerning metallothioneins (MTs) (proteins containing cysteine), which are able to link toxic metals [1].

The expression "actual state of health" we used in the first paragraph of this chapter aims at stressing the necessity of an evolution of scientific knowledge in the field of ecotoxicology and more importantly the necessity of employing alternative methods of environmental quality evaluation. These methods, such as biomonitoring, the use of biomarkers and biosensors in the environmental field, represent a sphere of important prospects of future developments.

2   Traditional monitoring and biomonitoring

Traditional monitoring methods, as valuable and unsurpassed as they are in the sphere of Environmental Chemistry, do have some shortcomings:

1. considerably high costs
2. methodological problems
3. problems with the release of contaminants on an intermittent basis
4. effects on biological species
5. numerous and extensive samplings

Table 1 reports traditional analytical techniques for the analysis of environmental pollutants [2].

Biomonitoring has significant advantages over traditional analysis of abiotic matrices (water, sediments). Besides providing information on the bioavailability of contaminants, it simplifies the chemical analysis, eliminating the problem of the assessment of very low levels of contaminants; it prevents the risk of misinterpretations caused by sudden fluctuations in the environmental parameters at the time of sampling; thus, providing a measurement over time of the level of environmental contamination; it does not require numerous, extensive, and prolonged samplings in the areas under study. All the above goes to show the importance of biomonitoring as a means to control environmental quality.

The use of cosmopolite organisms to assess pollution has developed notably during the last few decades. Such organisms assume environmental contaminants and may be used as indicators of the bioavailability of a given contaminant over time, allowing – in certain cases – comparison between contamination levels in geographically different areas [2].

It is in this context the Organisation for Economic Co-operation and Development (OECD) countries have taken many initiatives for examining potentially dangerous products by proposing general programs for the monitoring and evaluation of environmental impact [2–7].

From an ecotoxicological perspective, we can consider as contaminants or producers of environmental stress, all chemical compounds that are fundamentally released into the environment as a result of human activities, and which cause damage to living organisms [2, 8].

3   European Union legislative framework on chemicals [9]

The chemicals that can be released into the environment are more than 100,000. There is a considerable lack of knowledge regarding most of them. It is in this context and framework that the enactment of an important document of the European Commission, the White Paper – Strategy for a future Chemicals Policy [9] is situated.

To cope with the obvious inadequacy of the knowledge relating to the environmental and toxicological properties of many of the chemicals present on the European market, the strategy proposed by the White Paper provides for the

Table 1: Analytical techniques for the analysis of environmental pollutants [2].

Pollutant	Reference instrumental methods
SO_2	Flame photometric (FPD)
	Gas chromatography (GC)
	Spectrophotometric (pararosaniline wet chemical)
	Electrochemical
	Conductivity
	Gas-phase spectrophotometric
O_3	Chemiluminescent
	Electrochemical
	Spectrophotometric (potassium iodide reaction, wet chemical)
	Gas-phase spectrophotometric
NO_2	Chemiluminescent
	Spectrophotometric (azo-dye reaction wet chemical)
	Electrochemical
	Gas-phase spectrophotometric
	Conductivity
Fluorides	Potentiometric method
PAH	High resolution gas chromatography associated with mass spectrometry (HRGC/MS)
PCDD	High resolution gas chromatography associated with mass spectrometry (HRGC/MS)
PCDF	High resolution gas chromatography associated with mass spectrometry (HRGC/MS)
Metals	Atomic spectrometry
Chlorine and hydrochloric acid	Volumetric method; spectrophotometric analytical method
Phosphorus	Gas chromatography
	X-ray spectroscopy

adoption of a single system for "existing" chemicals and for "new" chemicals, called REACH (Registration, Evaluation, Authorisation of Chemicals).

This system provides for the acquiring of factual data on the chemicals, following three levels of inquiry in relation to the type and quantity of the chemicals concerned, involving increased responsibilities both for industry and for public authorities.

The EU legislation in force [10] provides for a clear differentiation between "existing" substances, or those put onto the market before 18th September 1981 and "new" ones put on the market subsequently. The latter (around 2,900) have had to pass a health and environmental risk assessment, based on the results of experimental tests supplied by the manufacturing companies.

The system of notifying "new" substances involves the obligation on the part of manufacturing or importing companies to present to the relevant national authorities a number of detailed items of information about the physico-chemical, toxicological, and environmental properties of the substances – information which has been specified in detail in the various directives that have several times modified the EEC/67/548 directive [11].

For the other substances, which were already on the market before 18th September 1981 – the 100,106 substances defined as "existing" and listed in the European Inventory of Existing Commercial Chemical Substances – the available information is, by contrast, still scarce if not actually non-existent. They represent 99% of the volume of the substances marketed (these are 100,106 substances, comprising industrial chemicals, substances obtained from metals, minerals, and other products present in nature such as petroleum, substances derived from animals and plants, food additives, active substances of pesticides, fertilisers, medicines and cosmetics, natural monomers and polymers, and some waste products or by-products).

Of these 100 thousand substances, according to data supplied by the industry, around 2,500 chemicals with a high volume of production (over 1,000 tons/year), 15,000–20,000 chemicals with a "low" volume of production (between 10 and 1,000 tons/year) and almost 80,000 chemicals in quantities lower than 10 tons/year are manufactured or imported on the European market.

The chemicals manufactured or imported in the EU in quantities above one ton/year (including those with a high volume of production) amount overall to around 30,000 and for only about a hundred of these – those considered "priority" according to the regulation (EEC) no. 793/93 – is a risk assessment program provided, involving manufacturers, national authorities, and the European commission.

The White Paper presented by the European Commission starts therefore from the observation that some tens of thousands of chemical substances are manufactured or imported on the European market, for which we do not yet know the toxicological and environmental properties.

The slowness of the assessment program for the "priority substances," at present under way, was criticized both in the White Paper and in the report which the Commission presented at the end of 1998 [12]; from this document, it emerges that only 19 assessment reports were completed by the end of 1998 out of over 100 substances indicated as priority and that the time needed from when a chemical is inserted in a priority list to the completion of the report on the conclusions of the assessment varies on average from two to four years.

If this slowness is due in part to the complexity of the examination required, as well as to the meagerness of the resources devoted to assessment activities, it is also true that the subdivision of responsibilities between manufacturing companies and national authorities does not facilitate the speeding-up of the program.

While in the case of "new" chemicals, industry has everything to gain from cooperating with national authorities, supplying all the necessary data so that the products may be assessed and marketed as quickly as possible, in the case of "existing"

chemicals that are already on sale, the onus of proof falls *de facto* on the public authorities with no specific obligations for companies about the data to be supplied in order
to allow an appropriate risk assessment. Proving an "unacceptable" risk for human
health and for the environment therefore rests entirely with the public authorities.

The EU is in the process of making the most fundamental changes to its legislation on the management of chemicals for over 30 years. Final political agreement
was reached at the Environment Council on 18th December 2006 after a severe
period of negotiations between the key stakeholders (the European Union, Member
States, the European Parliament, and in particular the Chair of its Environment
Committee, Directorates-General, Environment and Enterprise of the European
Commission, and others (i.e. industry, non-governmental organizations).

The final REACH text (more than 800 pages) was published in the Official
Journal of the European Union on 30 December 2006 and can be found at:
http://europa.eu.int/eur-lex/lex/JOHtml.do?uri=OJ:L:2006:396:SOM:EN:HTML
REACH entered into force on 1 June 2007; it now replaces over 40 existing
Directives and Regulations.

Another important issue is the one concerning the discussion of the problems
inherent in assessing the environmental risk of persistent organic pollutants (POPs),
which was the subject of a recent international convention in the United Nations
Environment Programme (UNEP) (Stockholm Convention on POPs, 2001) [13].

3.1 Categories of contaminants

The compounds released into the environment as a consequence of human activities can be classified in two main categories: biodegradable substances and
conservative substances [14].

Biodegradable substances represent the larger volume of human-generated
wastes (domestic wastes, industrial wastes, etc.) and, depending on their origin,
they are disposed of in the air, the soil, or water. They consist of organic material
(rich in C, N, and P) liable to bacterial degradation through oxidative processes
that reduce these organic compounds to soluble inorganic compounds (CO_2, H_2O,
and H_3N). If the release of these compounds is very high, anaerobic activity
occurs, with ensuing origination of degradation products (H_2S, HN, H_4C) that
not only give off unpleasant smells, but are also toxic for many organisms.

Generally, uncontrolled release of biodegradable waste products, especially
into water bodies, causes eutrophication phenomena and subsequently a decrease
in the quantity of oxygen present in the medium in question; it can also bring
about the production of toxic degradation compounds.

Conservative substances are not decomposed by bacteria or other short-term
processes. These substances are typically very reactive toward plants and animals,
sometimes causing considerable damages. There are three main categories:

1. heavy metals (Pb, Cd, Hg, Cr, Cu, Zn …);
2. halogenated hydrocarbons (HHC), dichlorodiphenyltrichloroethane (DDT),
   and polychlorinated biphenyls (PCBs);
3. radioactive compounds.

The components of these categories can be regulated by animals and plants to highly variable degrees, depending on the species, but always in an interval common to all species.

Halogenated compounds and metals are not eliminated; they accumulate over time in the tissues of the organism concerned and stay there permanently.

For instance, in aquatic systems, those organisms that are predators of bioaccumulators may have a diet rich in these conservative substances, which sometimes causes very high concentration levels of these compounds. Hence, there arises an exposure risk for trophically superior organisms (among which man) with subsequent problems concerning the release of these substances into the food chain [15].

4   Kinds of biomonitoring [2]

In general, *bioindicators* are organisms that can be used mostly for the identification and qualitative determination of human-generated environmental factors [15–17], while *biomonitors* are organisms mostly used for the quantitative determination of contaminants and can be classified as being *sensitive* or *accumulative*.

Sensitive biomonitors may be of the *optical* type and are used as integrators of the stress caused by contaminants, and as preventive alarm systems. They are based upon such optical effects as morphological changes in abundance behavior related to the environment or upon such chemical and physical aspects as alteration in the activity of different enzyme systems as well as in photosynthetic or respiratory activities.

Accumulative bioindicators have the ability to store contaminants in their tissues and are used for the integrated measurement of the concentration of such contaminants in the environment. Bioaccumulation is the result of the equilibrium process of biota compound intake/discharge from and into the surrounding environment. Bioaccumulation is the enrichment of a substance in the organisms through every pathway (respiration, food, dermal exposure).

The first studies of bioindicators date back to the 1960s. Beginning with the theoretical calculations of Stöcker [18] and Phillips [19, 20], we can define the main characteristics of a bioaccumulator.

Bioaccumulators must [2]:

- accumulate the pollutant without, however, being killed by the levels with which it comes into contact;
- have a wide geographical distribution;
- be abundant, sedentary, or of scarce mobility, as well as being representative of the collection area;
- be available all year round and allow for the collection of sufficient tissues for analysis;
- be easy to collect and resistant to laboratory conditions, as well as being usable in laboratory studies of contaminant absorption, if necessary;

- have a high *concentration factor* (CF) for the contaminant under study, and thus allow direct analysis with no prior increase in concentration;
- have a simple correlation between the quantity of contaminant contained in the organism and the average contaminant concentration in the surrounding environment;
- have the same contaminant content level correlation with the surrounding environment in every site studied and under any condition. This must be true for all organisms examined.

5   Reference levels, concentration factor, and biomagnification

An ecosystem must be understood as an entity that is relatively stable over time and having functional autonomy where there is an energy flow between organisms belonging to a complex trophic network.

The interactions that may occur among organisms are numerous and therefore also complex ones. When performing biomonitoring and traditional monitoring programs it is very important to consider these factors. Slobodkin [21] distinguished three different kinds of interaction among organisms:

1. Alterations of an indirect type that is when an organism alters the physical environment of another organism. For instance, the trees in a wood cast shadows and therefore reduce the photosynthetic activity of the surrounding plants;
2. Alterations (of an indirect type) of the physico-chemical environment, such as the increase in oxygen concentration, as a result of photosynthesis, in lake ecosystems on the part of unicellular algae. This oxygen is available for the respiratory activity of other organisms;
3. Exchange of chemical compounds, elements, or energy among organisms; many wild animals, for instance, constitute the food of both big and small predators, while the seeds of many plants provide several species of birds and mammals with energy, proteins, and vitamins (alterations of a direct type).

So, biomonitoring can be considered alongside traditional monitoring as it allows environmental quality assessment over wide geographical areas and an integrated measurement *over time* of the presence of contaminants.

For a variety of reasons, it is of fundamental importance to define the reference levels for pollutants in an ecosystem when making biological monitoring studies to:

1. evaluate the state of conservation or degradation;
2. predict the incidence of possible future human activities in order to establish the necessary interventions;
3. control evolution over time, using monitoring programs, if necessary.

To correctly evaluate the degree of contamination in an ecosystem, or to carry out biomonitoring operations, it is necessary to first establish the *background level* (BL) of the contaminant, both in the environment (air, water, soil), and in the organisms [15]. The BL may be interpreted in different ways: it may be understood as a *pre-industrial level* (prior to any human activity); as a *natural level* (the average conditions of an area or a region where there may be human activity, but which is in a good state of conservation); a *standard level* (based upon global geographical references); or even a *zero level* (the concentration of an element in the environment or in an organism prior to the development of a particular activity that is independent of the degree of conservation) [15].

Once the BL has been established, we use the CF to evaluate the state of conservation of an ecosystem, or to monitor its state. It is also called "contamination factor" mainly when this refers to biota/biota or abiotic compartment/abiotic compartment ratios of contaminants concentrations.

The CF is the relationship between the level of a contaminant found in the biota or environment and a reference value that represents a determined stage (pre-industrial, natural, zero):

$$CF_b = \frac{C_b}{BL_b} \quad \text{or} \quad CF_a = \frac{C_a}{BL_a}$$

CF = the contamination factor for the biota $(_b)$ or the environment (air, water, soil) $(_a)$; C = the concentration of contaminant in the biota $(_b)$ or in the environment $(_a)$, respectively; BL = the background level, of the pollutant in the biota $(_b)$ or in the environment $(_a)$, respectively.

If the BL is a reference of the zero level, it will allow us to observe the evolution of a pollutant (in terms of both space and time), during a contamination process.

In this case CF can be used as:

$$CF = \frac{Cb_{2A}}{Cb_{1A}}$$

Cb_{1A} = the concentration of pollutant present in the biota at the time 1 in the site A; Cb_{2A} = the concentration of pollutant present in the biota at the time 2 in the site A.

This concept may also be used to observe the decontamination rate in an ecosystem (positive impact). This signifies that the CF can be used also when, during comparisons between different environmental situations, no data is available for the BLs.

The CFs, in the monitoring programs, are also intended as the ratio between the concentration of the pollutant in the biota/BL of the contaminant in the environment at the same time of sampling:

$$CF_b = \frac{C_b}{BL_a}$$

For example, the Pb concentration measured in a mollusc tissue vs. the Pb concentration in the seawater soluble fraction. This reflects the concentration capabilities of the biomonitor with respect to the surrounding environment. It is generally correlated with the bioavailable fraction of the contaminant in the environment.

The system for environmental classification is realised by starting with the CFs obtained for each contaminant present in the environment or organisms. When evaluating the CFs obtained, it is also necessary to take into account the uncertainties that derive from the following: sampling; space, and time variations for the samples; the age and condition of the organisms, etc. In general, a CF that is above a given number (generally 1.5, 2, or 3 times the BL), is taken to be the minimum level under which it is no longer possible to refer to certain contamination. The qualification of a contamination situation may follow a linear scale, or, in high-level pollution conditions, a scale of the exponential type. Of course, the CF of each contaminant varies according to the species.

It is generally assumed that the release of the compound during the time unit is proportional to its concentration in the tissues. In these conditions, the higher the absorption constant of the contaminant and the lower the release constant, the higher the CF will be: the CF is the result of the processes of intake/discharge of the contaminant on the organism tissues. When using the CF as an indicator for the assessment of the quality of an ecosystem, we must take into account the possibility of self-regulation phenomena (detoxification) on the organism tissues which might alter the CF values.

The accumulation efficiency of an organism can be measured performing accumulation kinetics experiments, by keeping the contaminant concentration in water or air steady at levels that are not harmful to the organism itself, so as to determine the accumulation (or the mass growth) of the organism in terms of time (saturation curve). A contaminant can enter an organism through respiration (reversible process) or through food. Fig. 1 shows a general outline of the bioaccumulation phenomena.

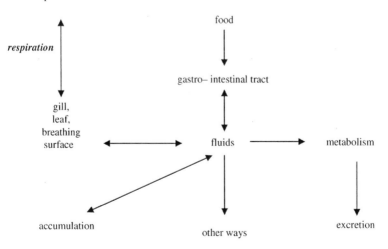

Figure 1: The outline of bioaccumulation phenomenon.

Bioaccumulation is thus understood as the intake of a contaminant through all possible pathways. *Bioconcentration*, on the other hand, is the direct intake of a contaminant exclusively through respiration.

The concentration of a number of compounds (as for instance mercury) accumulates every time it passes through the food chain. This phenomenon is known as *biomagnification* or concentrations along food chains.

However, biomagnification is a relatively rare phenomenon. One should demonstrate that the food pathway for the intake of a contaminant is clearly predominant over the others (especially respiration). Moreover, the accumulation of contaminant should increase with the passage from prey to predator, which is not always the case.

Some contaminants have much higher transfer efficiency than others: in these instances the losses through the passage from a compartment to another along the trophic chain are negligible. Such is the case of dichlorodiphenyldichloroethane (DDD), an insecticide of the DDT family, that was studied by Hunt and Bischoff in 1960 [22]. It was observed that the concentration of DDD increased as a consequence of the passage from water to aquatic organisms and from prey to predator. The concentration increased even more in the passage from fish to fish-eating birds (*Aechmophorus occidentalis*) situated at the top of the trophic chain.

In particular, the authors found that the DDD levels in the water of the Clear Lake, California were 0.02 ppm; in plankton 5.3 ppm ca.; in small fish 10 ppm; in predatory fish 1,500 ppm and in the western grebe fat 1,600 ppm. All this highlighted the biomagnification phenomenon.

The biomagnification of hydrophobic substances with a high resistance to degradation, such as DDT or PCBs, can be ascribed to the different physiology of the various organisms. If we consider that the air/water partition coefficients (K_{AW}) (at 20 °C) of DDT and PCBs are 10^{-2}–10^{-3} we can infer that their water concentration is 100–1,000 times higher than air concentration. Therefore, the transfer capacity of the contaminant from and to water organisms is two–three orders of magnitude higher than it is in air. As a matter of fact, aquatic organisms provided with gill have to extract the oxygen they need from water, when the oxygen concentration in water is 1/27 compared to air concentration (in equilibrium conditions at 20 °C). The gill must therefore work incessantly in order to extract a sufficient amount of dissolved oxygen from the water, which entails also a high exchange efficiency of the contaminants. For fish, the deposit compartment of contaminants is non-polar lipids that are easily assimilated by predators. This increases the possibility of biomagnification phenomena.

Therefore, assimilability is the decisive factor for biomagnification. We can observe that the levels of hexachlorobenzene (HCB), DDT, and PCBs in the grass of a meadow have values that can be compared to those present in fish, although the concentration of these contaminants are 100–1,000 times lower than those that can be found in the aquatic compartments. The above-mentioned contaminants accumulate in wax and possibly also in the cutin present in grass. These elements are not easily assimilated by ruminants, and therefore there is no biomagnification, due to the low transfer efficiency of the contaminant from the grass to the ruminant. The accumulation of contaminants in the waxes of terrestrial plants reduces sensibly biomagnification in herbivores [23].

Being lipophilic, contaminants often accumulate in the fat tissues of the organisms. The CF of an organism grows in proportion to the growth of K_{OW}, which is defined as the water/n-octanol partition coefficient. This partition coefficient is used because it is highly assimilable to lipids; it has a long hydrocarbon chain and has an alcoholic terminal function.

The measurement of the K_{OW} is considered to be very important, since several American, European, and Japanese environmental bureaus require them as indicators for the assessment of the quality of new compounds that are being put onto the market.

However, some recent works stress the fact that there has been a high level of uncertainty in reported octanol–water partition coefficients and aqueous solubilities (S_W) for DDT and dichlorodiphenyldichloroethylene (DDE) over the last six decades [24]. There was found four orders of magnitude variation in the K_{OW} and S_w database values for these hydrophobic organic compounds. The whole data quality seems to need to be reconsidered [24].

Biomagnification, however infrequent, can occur with methylmercury, the organic form of mercury found in water ecosystems, where it is very stable. Methylmercury is highly photolabile and is therefore found in places with scarce luminosity.

We all remember the tragic accident at Minamata Bay (Japan), where a chemical industry employing Hg^{2+} as a catalyst in the production process of polyvinyl chloride used to discharge wastes contaminated with mercury into the sea. Methylmercury bioaccumulated in fish, and fish was the staple food for the people living along the coast. The concentration of mercury in the fish was higher than 100 ppm (in the USA the recommended limit for mercury in fish for human food use is 0.5 ppm).

During 1950s, thousands of people living in Minamata suffered from mercury poisoning and hundreds of them died.

Another more recent disastrous accident is the one involving the petrochemical industry Enichem of Priolo, (Sicily): up to 2001, mercury used to be illegally disposed of in the sea. The levels of mercury measured in the sea were 20,000 times higher than the ones allowed. Congenital deformities in fetuses had been reported for some time in the area.

The K_{OW} of methylmercury is ca. 1, which means that its hydrophobicity is very low and its bioaccumulation capacity is virtually negligible. However, it penetrates cellular membranes easily (since it does not differentiate lipids from water, as indicated by its $K_{OW} \approx 1$) thus forming stable complexes with thiolic groups of the proteins inside the cells where it builds up.

Transfer efficiency from a trophic level to the next can be measured by the trophic transfer coefficient (TTC), which is the ratio between the concentration of the contaminant in the tissues of the consumer and the concentration in the tissues of the food (prey) [25]. Therefore, if the TTC value is lower or equal to one, there occurs no biomagnification phenomenon. If it is higher than one, there is biomagnification. Suedel *et al.* [25] detected TTC values as high as 100 for methylmercury (full-blown biomagnification). For several organic compounds studied (atrazine, chloro-dioxines, polycyclic aromatic hydrocarbons (PAHs)) the tests for biomagnification had negative results (TTC ≤ 1). Other contaminants (DDT, DDE, and PCBs) show TTC values ranging between 0.1 and 10. PCBs (highly hydrophobic) have TTC values as high as 10 in the passage from small to big fish, and it reaches an enrichment factor of 100 in the passage to dolphins and fish-eating birds [26, 27].

Another example of accumulation in the trophic chain is the one concerning radioactive depositions over the Mediterranean area after April 1986 (Chernobyl accident). The rapid removal of Cs-137 from water by biological activity has transferred this radionuclide to the pelagic and benthic communities [28]. Transfer of Cs-137 along the marine trophic chain in the Po delta (Adriatic sea) was found in 1990. Levels of Cs-137 (Bq/kg wet weight) were: 0.07 for plankton; 0.5 for *Merluccius merluccius*, and 0.6 for *Sardina pilchardus* [29].

6   Sampling problems

Before the assessment of the BL of contamination of an ecosystem, it is necessary to clearly define the area under study and the area that can be affected by a possible source of impact.

The area of influence depends on several factors:

1. the magnitude of the impact (e.g. volume of the pollutants released in the atmosphere);
2. the characteristics of the receiving body (e.g. wind direction);
3. the time scale that will be used.

The area of influence of a group of contaminants varies depending with time. Another problem is the representativeness of the samples and the intensity of the sampling. When establishing the number of samples, it is necessary to consider the size of the area of influence and the actual presence of the species under study in the area concerned.

The necessary number of samples changes with each kind of material that is analyzed (atmosphere, sediments, biota) and with each contaminant studied, since they have different affinities both regarding their bioconcentration capacity and their distribution in the ecosystem. As a general rule, the less we know about a contaminant, the more samples we need.

Bros and Cowell [30] developed a technique that employs the standard error (SE) of the mean to resolve statistical power in order to determine the number of samples needed based on number of species, number of individuals, biomass, diversity, and evenness. This method uses a Monte Carlo simulation procedure to generate a range of sample sizes vs. power. As the number of samples is increased, the rate of increase of SE declines, and eventually becomes near to zero level; the point where this occurs corresponds to the minimum number of samples required. Generally, the less abundant the pollutant, the greater the number of samples required to assess its level.

When quantifying the BL of a contaminant in a given organism, we refer to the mean concentration of the contaminant in the tissues of the organism. It is important to take into account the condition index (age, sex and maturity, size, etc.) of the population under study. These elements can alter significantly the concentration of the pollutant in the organism. For example, in molluscs the concentration of the contaminant depends on the weight (size). Therefore, in biomonitoring studies using molluscs, the concentration/weight factor is determinant and important for the processing of data (see further on).

The BL can also refer to the natural concentration of the pollutant in a given organ (brain, liver, pancreas, thallus, etc.) or to an analytical fraction that corresponds to a given extraction process (intra-extracellular, lipid fraction, etc.). This is because those organs, beside having a particular bioaccumulation ability, present less difficulties when chemical analyses are performed; or also for the different biological meaning that this localized bioaccumulation takes on. The data obtained this way are generally more stable than those obtained from the analysis of the whole organism.

6.1 Sample collection

The methods and materials used to collect, store, and transport samples should be considered carefully. Sampling devices and their materials of construction should have to be evaluated under specified conditions. Contamination by sampling devices and materials can contribute relatively large errors in comparison to analytical procedure, especially when the analytes of interest are at low concentrations. Sampling protocol should be based on labile analytes to be measured. The term labile is hence regarded as the metal forms with the highest probability that the analyte concentration change prior to analysis.

Composite sampling defined as combining portions of multiple samples is of advantage in a monitoring program. Even when each individual test sample continues to be the material of study used in many biomonitoring studies composite sampling frequently replaces collection of individual specimens [31].

Composite samples are often used to reduce the cost of analyzing a large number of samples and also to diminish intersample variance due to heterogeneity of the sampled material. Statistical evaluation of the results obtained in composite samples indicate less mean square error in the frequency of analyte occurrence than in the approach of sampling and analyzing one individual from each one of the multiple populations collected.

Another advantage of composite sampling is that it may also increase the amount of sample available for analysis, especially when each individual furnishes too little quantities of material of study.

After sample collection, appropriate conditioning and storage precautions are of concern in order to minimize the risks of analyte loss or contamination.

For speciation studies of biological materials more stringent conditions have to be fulfilled during sample collection, pretreatment, and storage. It is of relevance to maintain the integrity of the chemical species during and after sampling. Changes in parameters such as temperature, ionic strength, pH, redox potential, oxygen level, irradiation with UV light, etc., to which the samples are exposed, can influence the distribution of chemical species.

The choice of sample storage containers is also a critical factor, being deep freezing samples an appropriate technique to perform immediately after collection to minimize bacterial or enzyme degradation.

6.2 Sample preparation

Until the middle of the last decade sample preparation procedures for biological materials fell into two categories. Firstly, analytical methods with minimum sample treatment (nearly intact sample) and secondly, those methods where an important sample treatment such as separation, extraction, or destruction of the organic matter is necessary before determinations can be carried out.

Sample preparation in inorganic analytical chemistry is generally concerned with digestions, which principally serve to liberate target compounds from the sample matrix and convert the various chemical forms of the analyte to a uniform species [31].

Analysis of plant materials and animal tissues usually involves destruction of organic matter. Wet acid and dry ashing techniques are commonly used for this purpose but the former has been gaining more acceptance during the last years. Digestion by refluxing in boiling concentrated acids such as nitric acid or nitric/ perchloric acids is the conventional technique used in biomonitoring studies but special microwave heating for acid digestions has lastly increased the speed and efficiency of the operation (for more details on digestion procedures see Chapter 5).

Sample preparation in organic analytical chemistry mainly comprises extractions, which serve to isolate components of interest from a sample matrix because most analytical instrumentation cannot handle the matrix. Extraction techniques using large amounts of organic solvents in analyte laboratories such as the Soxhlet extraction method are still used. However, several disadvantages of solvent extraction in routine analysis are known: sample preparation is time consuming, employ multi-step procedures with high risk of analyte loss and are rather expensive [31].

In biomonitoring the main drawback of the entire process would probably be that most of the analysis time is consumed by sampling and sample preparation, mainly because of the large number of samples to handle and process. Microwave assisted extraction (MAE) techniques are an interesting alternative for separation of organometallic compounds in environmental and biological samples. Extraction in an open microwave oven is of great advantage for dissolving samples because it offers reducing extraction time and reagents, thus reducing contamination problems. Furthermore, procedures for sequential extraction using microwave heating under controlled conditions were also established [31].

In the last years a novel technique, solid phase microextraction (SPME), regarded as a sorbent extraction technique and a solvent-free sample preparation, has been applied for trace element speciation in environmental and biological samples. One of the main advantages of SPME in metal speciation is that it is a non-exhaustive technique permitting equilibrium of the target compound between the free and bound-to-matrix forms be practically undisturbed during the extraction procedure [31].

Volatile and non-volatile organometallic compounds can be collected by fibers placed into a tip of the SPME system at the entry of gas chromatography (GC) and liquid chromatography (LC) instruments, respectively, used for their separation. Only a soft digestion of the sample to drive the analytes into the liquid phase is needed for immersion of the tip, though in other cases the fiber is in contact with the gas phase of the headspace in the chamber containing the sample solution [31].

6.3   The use of certified reference materials

National and international marine monitoring programs have been initiated world-wide to assess the quality of the marine environment. In environmental analysis *Reference Materials* (RMs) play an important role to achieve quality in the results; therefore, an appropriate use of reference standards must be emphasized.

Definitions of some useful terms related to RMs are given in ISO Guides [32–34]. RM is a material or substance one or more of whose property values are sufficiently homogeneous and well established to be used for the calibration of an apparatus, the assessment of a measurement method, or for assigning values to materials whereas a *Certified Reference Material* (CRM) is a Reference Material, accompanied by a certificate, one or more of whose property values are certified by a procedure which establishes its traceability to an accurate realization of the unit in which the property values are expressed, and for which each certified value is accompanied by an uncertainty at a stated level of confidence.

Traceability implies that a measurement result should be related to stated references, and therefore, the value reported in the certificate of a CRM is traceable. So, when CRM are used in a method the user has to demonstrate its traceability. *Traceability* is defined as a property of the result of a measurement or the value of a standard whereby it can be related, with a stated uncertainty, to stated references, usually national or international standards, through an unbroken chain of comparisons.

Certification of a Reference Material is a procedure that establishes the value(s) of one or more properties of a material or substance by a process ensuring traceability to an accurate realization of the units in which the property values are expressed, and that leads to the issuance of a certificate.

The large array of matrices encountered in the environment makes necessary the use of a very wide variety of RMs derived from different sources, related to the intended purpose of the user. Advantages and choices of CRM as well as the procedure for carrying out the measurement are described in the International Organization for Standardization (ISO) Guides [32–34].

CRM chosen must adequately match the material to be analyzed, and it also implies that the levels of the analytes or target compounds have to be similar to those of the real samples used for analysis. Traceability of CRM will be questionable when the matrix and analyte contents are too different from those of the analyzed sample.

The contents of trace, minor, and major elements in CRM for the quality control of marine and estuarine monitoring programs include biological, sediment, and seawater samples. Technical information and purchasing conditions are provided from the main suppliers by request.

Usually, most elements exert their biological and environmental impact as components of macromolecules, or linked to small organic compounds, or according to a specific oxidation state. Therefore, not only total element determinations are of interest any longer but also identification and quantification of element species (speciation).

In the last twenty years, speciation analysis in the environmental and life sciences has mainly focused on the analysis of relatively stable organometallic compounds such as some organic forms of mercury, tin, or arsenic, and oxidation states: Cr (III)/Cr(VI).

Today, there is a great interest in speciation studies of many other elements present in the marine environment and biota. It is also necessary that availability, and distribution, transport, transformation, and fate of chemical species of elements regarded as toxic, rather than of the total amount of the elements, be studied in very complex systems and samples taken from biotic and abiotic sources. Unfortunately, there is a limited number of available CRM to fit these purposes. Existing CRM are intended to be used for trace element speciation and fractionation purposes. Explanation of these two fundamental concepts has been extensively treated in a publication of the International Union of Pure and Applied Chemistry (IUPAC) [35].

7 Determination of background levels (BLs)

The determination of BLs [15] is fundamental to determine the degree of enrichment of the contaminants in the environment or in the biota. Generally, the data used as BLs come from two different sources:

1. mean values used as general reference levels
2. values determined in their area of study

Both procedures are valid and the choice between them depends on the goal one is trying to achieve. Using general reference levels local variations will be ignored. On the other hand, using the pre-industrial levels of the areas concerned, local variations will actually be emphasized.

However, for an exact environmental assessment and/or monitoring of a given geographical area, it is generally preferred to use the values determined in the specific area of study, even though this entails higher costs, sampling and laboratory work. In this approach, it is important to minimize the effect of the variability of the samples in the selected sites.

The difficulty with these methods of approach to study lies in the ability to distinguish between the variability due to contamination and natural variability, as well as to report data on a regional scale. The key for a correct definition of BL is to select unpolluted sites. This selection can be performed by choosing:

1. for each contaminant the sites that are not polluted by this particular contaminant;
2. the sites those are also free from contamination of typically polluting elements.

When selecting uncontaminated stations, before sampling a geographical area, it is necessary to follow a number of preliminary procedures [15].

In accordance with the information available, a series of stations (or sites) can be selected that are considered representative of natural levels. This decision entails a high degree of risk if there are no exhaustive preliminary data on the area of study.

The selection of the sites can be made using multifactorial techniques. With this method, clean stations are selected after being grouped through multifactorial analysis (principal component analysis (PCA), dendrograms, etc.) applied to a matrix of data that are linked to the quality of each ecological station.

Elements of segregation can be used, together or separately, such as the physico-chemical parameters of the environment, the concentration of the pollutants in the environment and organisms, the frequency of the species of bioindicators, etc.

However, this method presents some problems. The results can be difficult to interpret and may have low grouping levels, especially if the majority of the samples have low levels of contamination. Furthermore, contamination is distributed along different gradients that depend on the natural variability of the physico-chemical conditions of the environment (air, soil, water). This method is more effective when the areas of study present more homogeneous environmental conditions and the contamination points are more punctual and intense.

Another method for the selection of clean sites involves bioindicators. Ecological stations are selected on the basis of a possible manifestation of stress in one or several species of environmental quality indicators.

The stress to be measured may be biochemical, physiological, or structural (e.g. chlorophyll degradation). Usually, several bioindicators are used belonging to different biological species present in that environmental compartment and responding to pollutants in a complementary way. This method allows the evaluation of the degree of natural preservation of the selected sites with a reasonable degree of approximation.

When the contaminant type is known, some indicators can be used more effectively and this allows the use of less costly methods. If the contaminants are not known and the information on the tolerance level of the species is scarce, a multiple survey and possibly toxicity tests are necessary.

A multiple level survey entails the study of possible indicators and changes in the behavior of the organisms. The indicators have to show biochemical, genetic, morphological, or physiological changes (biomarkers). The behavior indexes are determined by the changes of given species, dynamic populations, or communities (e.g. freshwater macroinvertebrates).

The biomonitoring of communities provides information on the magnitude and the ecological effects of the contaminants in an ecosystem. The cause/effect relationships are difficult to establish and the knowledge in this respect is still dubious, since there are many factors affecting the community-contaminants system. Therefore, the use of bioindicators at different organizational levels (for instance: individuals, species, communities, ecosystems) is more advisable.

If the choice of the uncontaminated stations concerns a given contaminant, there are different statistical methods that can be used, such as the use of data populations with a coefficient of variation (CV) of approximately 60%. This method is used for the assessment of the natural geochemical BL in soils and consists on the progressive elimination of the higher values of the contaminant under study until a CV of 60% is reached. Such population distribution is considered normal [14, 15].

Many ecotoxicological studies use the determination of homogeneous populations within a number of data belonging to a given site, employing graphs and diagrams. These studies rely on the analysis of distribution curves of cumulative frequencies transformed in a log-normal distribution (cumulative frequency curves). It is thus possible to distinguish homogeneous populations (straight lines in the diagram) that correspond with the different contamination levels: base level, middle level, and contaminated. The problem with this method is that it is necessary to have a significant amount of data in order to have a valid graphical interpretation of the populations.

Another method involves normal populations. In order to solve the problem of size in the number of data – which is a common problem in most works, modal analysis methods are also applied. Such methods are often used in demography studies employing adequate software. For instance, NORMSEP (SEParation of the NORMally distributed components of size-frequency samples) is a program that transforms the frequency of the concentrations obtained into a distribution of normal components, thus allowing a differentiation between the various homogeneous populations that may coexist with the complete series of data. Thanks to this method it is possible to work even with a reduced number of samples.

The use of regression techniques is applied between a stable element that is not affected by anthropic activity and the rest of the contaminants. In this instance, the clean stations will be those presenting a confidence level of 95% of the regression line. However, the use of this approach can contribute to eliminate some uncertainties as to a clear separation between clean and contaminated sites. In view of this, in order to have a statistically meaningful regression, it is necessary to have a high number of samples [36, 37].

The systematics of environmental classification is obtained starting from the CFs obtained for each contaminant in the environment or in the organisms.

When evaluating CFs, it is important to consider all uncertainties due to sampling, space, and time variability of the samples, age, and condition index of the organisms, etc.

The various elements of the uncertainty of measurement in the different steps (sampling, pretreatment, analytical steps) must be duly evaluated within the framework of the approach that has been adopted for the study.

Generally, two approaches are used for the evaluation of uncertainty:

1. the theoretical "bottom-up" approach recommended by international organizations. This method requires the evaluation, expressed as standard deviations, of all factors that will contribute to the final value (e.g. volumetric flask corrections, standard weight corrections, pipet volume corrections, signal uncertainty, etc.).
2. the practical "top-down" approach from the relative standard deviation derived from an interlaboratory study by the Harmonized IUPAC/Association of Official Analytical Chemists (AOAC) protocol [38].

The formal definition of uncertainty given by the ISO Guide [39] is as follows: "Parameter associated with the result of a measurement, that characterizes the dispersion of the values that could reasonably be attributed to the measurand."

Recently, Conti *et al.* [38] analyzed theoretical and practical aspects of uncertainty in environmental laboratory analysis.

In general, a minimum level, below which it is not possible to talk about a certain contamination, is a CF that is higher than a given amount (generally 1.5, 2, or 3 times higher than BL). The qualification of a situation of contamination can follow a linear scale. However, it can also happen that, in the presence of a high level of pollution, there is an exponential scale. Table 2, for instance, shows the contamination situation that can occurs in relation to the CFs of estuary ecosystems.

This is a useful means to control the impact basal levels of the different ecosystems. It also allows a correct formulation of environmental prevention programs both on a large scale and on an industrial level, like in the Environmental Impact Assessment (EIA) studies. As for instance before the building or modification of

Table 2: Concentration factors and contamination of estuary ecosystems.

CF	Situation
<2	negligible
2–6	possible-moderate
6–18	certain-severe
18–54	very severe
>54	critical

big industrial plants (refineries, thermal power plants, steelworks, etc.) or large infrastructure works (public buildings, motorways, etc.).

8   Levels of organization

In short, in order for an organism to be considered a valid bioindicator, it must have a wide geographical distribution, be available all year round, and be also very tolerant to the contaminant (without being killed by it), be easy to sample and store, and, more importantly, it must have a positive correlation between the contaminant concentration accumulated in its organism and the concentration of that same contaminant in the surrounding environment [2].

Table 3: Organization levels and measures connected with biomonitoring studies [15].

-individual
-organism
-genetic mutations
-reproductive success
-physiology
-metabolism
-oxygen consumption, photosynthesis rate
-enzyme/protein activation/inactivation
-hormones
-growth and develop
-resistance to disease
-tissue/organ damage
-bioaccumulation
-population
-survival/mortality
-sexual rate
-abundance/biomass
-behavior (migration)
-predation rate
-decrease/increase population
-community
-organism (or organisms) abundance
-biomass
-organism (or organisms) density
-abundance (variety), number of species, width class, or other functional
-groups, per area or volume, or number of individuals
-variety/relative abundance of species
-ecosystem
-mass nutrients

The identification of cosmopolite indicators allows a comparison between the absolute levels of some contaminants (e.g. metals) of organisms belonging to different geographical areas. Bioindicators are a very valid tool because they can accumulate contaminants such as heavy metals from the aqueous medium up to tens of thousands of times as much as BLs.

Measurements associated with biomonitoring are possible at various levels of organization: individual, population, community, and ecosystem. A synthesis of these by now commonly used methods is reported in table 3.

The identification of cosmopolite indicators allows a comparison between the absolute levels of some contaminants (e.g. metals) of organisms belonging to different geographical areas. Bioindicators are a very valid tool because they can accumulate contaminants such as heavy metals from the aqueous medium up to tens of thousands of times as much as BL.

References

[1]   Conti M.E. (2002) Gli indicatori ed i marcatori biologici in Ecotossicologia. Università degli Studi di Roma, in: Facoltà di Farmacia (Ed.), La Sapienza. Centro Stampa d'Ateneo, Rome, pp. 124–141.

[2]   Conti M.E., Cecchetti G. (2001) Biological monitoring: lichens as bioindicators of air pollution assessment – a review. Environmental Pollution, 114, 471–492, Reprinted (or higher parts taken) with a kind permission of Elsevier.

[3]   Connell D. (1999) Introduction to Ecotoxicology. Blackwell, Oxford, UK, pp. 184.

[4]   Tessier A., Campbell P.G.C., Bisson M. (1980) Trace metal speciation in the Yamaska and St. François Rivers (Quebec). Canada Journal Earth Science, 17, 90–105.

[5]   Krumgalz B.S. (1989) Unusual grain size effect on trace metals and organic matter in contaminated sediments. Marine Pollution Bulletin, 20(12), 608–611.

[6]   Herman R. (1987) Environmental transfer of some organic micropollutants. Ecological Studies, 61, 68–99.

[7]   Bero A.S., Gibbs R.J. (1990) Mechanisms of pollutant transport in the Hudson Estuary. The Science of the Total Environment, 97/98, 9–22.

[8]   Moriarty F. (1999) Ecotoxicology: The Study of Pollutants in Ecosystems. Academic Press, London, pp. 350.

[9]   Zaghi C., Conti M.E., Cecchetti G. (2002) White paper on chemicals and Stockholm convention on persistent organic pollutants: perspectives for environmental risk management. International Journal of Risk Assessment and Management, 3(2–3–4), 234–245, Reprinted (or higher parts taken) with a kind permission of Inderscience Publishers.

[10] European Commission White Paper (2001) Strategy for a Future Chemicals Policy, COM, 88 final.

[11] Directive 67/548/EEC relating to the classification, packaging and labelling of dangerous substances as amended, directive 88/379/EEC and new directive 99/45/EC relating to the classification of dangerous preparations, directive 76/769/EEC relating to restrictions on the marketing and use of certain dangerous substances and preparations, regulation (EEC) 793/93 on evaluation and control of risks of existing substances.

[12] European Commission Working Document SEC (1998)1986 final.

[13] Stockholm Convention on Persistent Organic Pollutants (2001) URL: http://www.cham.unep.ch/pops.

[14] Carballeira A., Carral E., Puente X., Villares R. (2000) Extensive control of litoral contamination. Nutrients and heavy metals in sediments and organisms on Galicia coast (NW Spain), in: Conti M.E., Botrè F. (Eds.), The control of marine pollution: current status and future trends. International Journal of Environment and Pollution, 1–6, 534–572.

[15] Conti M.E. (2002) Il monitoraggio biologico della qualità ambientale. ed. Seam, Rome, p. 185.

[16] Hertz J. (1991) Bioindicators for monitoring heavy metals in the environment, in: Merian E. (Ed.), Metals and their Compounds in the Environment, VCH, Weinheim, 221–232.

[17] Tonneijk A.E.G., Posthumus A.C. (1987) Use of indicator plants for biological monitoring of effects of air pollution: the Dutch approach. VDI Ber, 609, 205–216.

[18] Stöcker G. (1980) Zu einigen theoretischen und metodischen. Aspekten der Bioindikation, in: Schubert R., Schuh J. (Eds.), Methodisce und theoretische grundlangen der bioindikation (Bioindikation 1), Martin-Luther-Universität, Halle (Saale), GDR, pp. 10–21.

[19] Phillips D.J.H. (1977) The use of biological indicator organisms to monitor trace metal pollution in marine and estuarine environments. A review. Environmental Pollution, 13, 281–317.

[20] Phillips D.J.H. (1980) Quantitative Aquatic Biological Indicators: Their Use to Monitor Trace Metal and Organochlorine Pollution. Applied Science Publishers, London, p. 488.

[21] Slobodkin L. (1961) Growth and Regulation of Animal Populations. Holt, Rinehart & Winston, New York.

[22] Hunt E.G., Bischoff A.I. (1960) Inimical effects on wildlife of periodic DDD application to Clear Lake. California Fish Game, 46(1), 91–106.

[23] Bacci E., Gaggi C. (1998) Bioconcentrazione, bioaccumuloe biomagnificazione, in: Vighi M., Bacci E. (Eds.), Ecotossicologia. UTET, Torino, pp. 143–152.

[24] Renner R. (2002) The K_{ow} controversy. Environmental Science and Technology, 36(21), 411A–413A.

[25] Suedel B.C., Boraczek J.A., Peddicord R.K., Clifford P.A., Dillon T.M. (1994). Trophic transfer and biomagnification potential of contaminants in aquatic ecosystems. Reviews of Environmental Contamination and Toxicology, 136, 21–89.

[26] Focardi S., Leonzio C., Fossi C., Rosi C., Marsili L. (1991) Delfinidi spiaggiati lungo le coste toscane e laziali: livelli di idrocarburi clorurati. S.IT.E. Atti, 12, 197–200.

[27] Focardi S., Marsili L., Fabbri F., Carlini R. (1990) Preliminary study of chlorinated hydrocarbon levels in Cetacea stranded along the Tyrrhenian coast of Latium (Central Italy), in: Evans P.G.H., Aguilar A., Smeenk C. (Eds.), European Research on Cetaceans, Cambridge 4, pp. 108–110.

[28] Fowler S.W., Buat-Menard P., Yokoyama Y., Ballestra S., Holm E., Nguyen H.V. (1987) Rapid removal of Chernobyl fallout from Mediterranean surface waters by biological activity. Nature, 329(6134), 56–58.

[29] Nonnis Marzano F., Triulzi C. (2000) Evolution of radiocontamination in the Mediterranean Sea in the period 1985–1995. International Journal of Environment and Pollution, 13(1–6), 608–616.

[30] Bros W.E., Cowell B.C. (1987) A technique for optimizing sample size (replication). Journal Experimental Marine Biology Ecology, 144, 63–71.

[31] Conti M.E., Tudino M.B., Muse J.O., Cecchetti G.F. (2002) Biomonitoring of heavy metals and their species in the marine environment: the contribution of atomic absorption spectroscopy and inductively coupled plasma spectroscopy. Research Trends in Applied Spectroscopy, 4, 295–324, Reprinted (or higher parts taken) with a kind permission of Research Trends.

[32] International Organization for Standardization – ISO Guide 33. (2000) Uses of certified reference materials.

[33] International Organization for Standardization – ISO Guide 35. (2006) Reference materials – General and statistical principles for certification.

[34] International Organization for Standardization – ISO Guide 32. (1997) Calibration in analytical chemistry using certified reference materials.

[35] Templeton D.M., Ariese F., Cornelis R., Danielsson L.-G., Muntau H., Van Leeuwen H.P., Łobiński R. (2000) Guidelines for terms related to chemical speciation and fractionation of elements. Definitions, structural aspects, and methodological approaches (IUPAC recommendations). International Union of Pure and Applied Chemistry, 72(8), 1453–1470.

[36] Conti M.E., Cucina D., Mecozzi M. (2007) Regression analysis model applied to biomonitoring studies. Environmental Modeling Assessment DOI: 10.1007/s10666-007-9113-7 (in press).

[37] Conti M.E., Iacobucci M., Cecchetti G., Alimonti A. (2007) Influence of weight on the content of heavy metals in tissues of Mytilus galloprovincialis (Lamarck, 1819): a forecast model. Environmental Monitoring Assessment DOI: 10.1007/s10661-007-9875-z (in press).

[38] Conti M.E., Muse J.O., Mecozzi M. (2005) Uncertainty in environmental analysis theory and laboratory studies. International Journal of Risk Assessment and Management, 5(2–4), 311–335.

[39] Guide to the expression of uncertainty in measurement (GUM). ISO, Geneva, 1993 (Reprinted 1995). Italian version: Ente Nazionale Italiano di Unificazione (UNI). UNI CEI ENV 13005. Guida all'espressione dell'incertezza di misura, Milano, UNI, 2000.

2 Biomarkers for environmental monitoring

M E Conti

1 Introduction

Many researches about toxicology, especially aquatic toxicology, have focused extensively on the early phases of the impact of a contaminant with an organism, characterized by its interaction with the endogenous molecules. Fig. 1 reports a synthesis of the interactions between the contaminants present in the aquatic ecosystem and their biological effects.

Toxicology evaluations do not have recourse to biological indicators in the strict sense of the term. In the field of toxicology, there are very interesting prospects for biomarkers, which are specific biochemical, genetic, morphological or physiological changes measurable in each organism and which are associated with particular stress situations (for instance, in the presence of heavy metals, pesticides, etc.). Biomarkers signal the occurrence of toxicological events much earlier than the emergence of those effects that can be measured in biocoenosis at the population level. The measurements of the biological effect ensuing exposure to contaminants follow two main criteria [1–4]:

1. The biological change must be caused *exclusively* by the contaminant. The evaluation must also takes into account the variables those are present all the time (for instance, season, gender, temperature) so as to establish a good signal/noise ratio.
2. It must also be correlated with a negative effect on some physiological aspects of the organism under study, such as growth, reproduction or survival. In particular, response times must be short, within the range of hours–weeks.

Depledge [5] proposed a more exhaustive definition of biomarkers. A biomarker is defined as a change at a biochemical, cellular, physiological or behavioural level; it can be measured in tissues and/or cellular fluids and/or in the whole organism and shows the exposure and/or the effects of one or more chemical contaminants (and/or radiations).

As briefly hinted above, the principal characteristic of biomarkers is the fact that they signal change events (also with localized toxicity effects) long before the appearance of measurable effects in the biocoenosis at the population level.

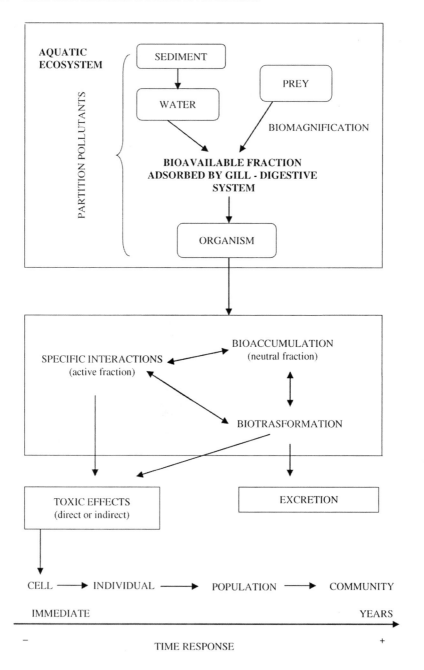

Figure 1: Interactions between pollutants and their biological effects in the aquatic environment (modified from Narbonne and Michel, 1993 [1]).

Biomarkers are used as early warning systems signalling potentially compromised situations. Preferably, the biological response should occur in the interval between the ideal conditions (beginning of the anthropic event) and the onset of lethal conditions of the organism under observation.

One of the shortcomings of the ecotoxicological approach is perhaps that it fails to take into account also the impacts of environmental xenobiotics on population dynamics, community structure and ecosystem processes [6]. Therefore, the measurement of single functions can be of very little significance if it is not related to measurements of Darwinian fitness such as growth rate, reproductive output and offspring viability [5].

2   Potential biomarkers

Biomarkers are classified as follows: exposure biomarkers and effect biomarkers. Exposure biomarkers are the responses of an organism, at its different levels of structural organization, to the exposure to a chemical compound or a set of chemical compounds. However, these biomarkers do not yield any information on the toxicological effects of the organism under study and are therefore used as an early warning system vis-à-vis a polluting event.

Among this kind of biomarkers there is for instance the inhibition of plasma esterases (butyrylcholinesterase (BchE) and carboxylesterase (CbE)) caused by organophosphorous insecticides. The inhibition of these enzymes does not cause any harmful effects in the whole organism.

Effect biomarkers are responses correlated both with the exposure to a contaminant and with its toxic effect. For instance, the effect of organophosphorous (OPs) insecticides and carbamates (CBs) triggers off the inhibition of acetylcholinesterase (AchE) which causes serious damages in the functioning of the central nervous system of many organisms.

Biomarkers can also be classified as general and specific. General biomarkers are all those responses of an organism at various levels (genetic, molecular, cellular, physiological and behavioural) which are not caused exclusively by one class of contaminants. These responses represent a *stress* situation of the organisms in the ecosystem studied.

Specific biomarkers are the molecular and biochemical responses detectable in organisms as a consequence of exposure to a particular class of contaminants. The following is a list of the most commonly studied biomarkers.

2.1   DNA or mRNA alterations

It is possible to utilize as biomarkers DNA alterations occurring at different structural levels. In this respect, several techniques can be developed in order to observe changes in gene expression as a result of exposure to some chemical stressors [7, 8].

Some techniques, such as mRNA "finger-printing", facilitate the identification of variability in gene expression for the whole genome [9]. This technique can be utilized in the field of environmental toxicology in order to verify the presence of contaminants with an unknown mode of action. It is based on the systematic amplification of the 3′ terminal portions of mRNAs and resolution of those fragments on a DNA sequencing gel thus observing altered functions in mRNA transcription or degradation rates [9, 10]. However, it is time consuming, labour intensive and suffers from a lack of specificity [10].

Other techniques verify the induction of specific genes as a response to particular classes of contaminants. For example, a molecular based assay has been developed to measure the induction of gene expression in sheepshead minnows (*Cyprinodon variegatus*) exposed *in vivo* to xenoestrogens (i.e. 17β-estradiol, ethinylestradiol, diethylstilbestrol) [11]. Results of this study demonstrate a characteristic expression pattern for genes upregulated by exposure to a variety of these contaminants.

Many contaminants (PAH, dioxins, etc.) are able to modify the genetic material. Some molecules (such as benzopyrene) bind steadily to the double helix thus forming adducts, which are reactive chemical intermediates that covalently bind to DNA bases. Adducts measure immediate responses in relation to the polluting event. Genotoxicants have the ability to alter DNA and their effects may be harmful because these contaminants can induce changes on to future generations [12]. In order to identify them, the immunochemical method ELISA (enzyme-linked immunosorbent assay) and the radiochemical method (^{32}P-postlabelling technique) [13] are employed.

Adducts can also cause fractures of the double helix thus provoking structural alterations that are easily quantifiable (secondary alterations). These fractures can always be traced back to immediate responses. When secondary alterations are significant and the organism has lost all capacity to recover there is a situation of irreversibility (chromosomal aberrations). In this instance, some cytologic tests are used to evaluate biomarkers, such as the micronucleus test, chromosomal analysis and Sister Chromatid Exchange (SCE), in order to assess the damage done.

If the damage reaches the mutation level, the oncogene activation method can be employed in order to verify the event.

2.2   Protein responses

These are the reponses of an organism following exposure to several classes of contaminants that are therefore characterized by induction (a) or inhibition (b) of the activity of functional proteins.

1. This group comprises the adaptive or protective mechanisms involved in the detoxification of xenobiotic compounds (mixed function oxygenase (MFO) system, conjugating enzymes) and detoxification mechanisms with respect to heavy metals. These are very specific biomarkers from a qualitative point of view and can also serve as a semi-quantitative signal of the presence of a particular kind of contaminants.

2. These are those biomarkers linked with phenomena of inhibition of the enzymatic activities (i.e. inhibition of the blood or brain esterases caused by organophosphorous insecticides).
3. Other general biomarkers are the induction of stress proteins and plasma proteins.

The MFO system is of great importance in the initial phase of detoxification of xenobiotic compounds (Phase I). This phase entails the creating or the modifying of the functional groups of the xenobiotic molecule. The functional centre of these enzymes is the cytochrome P450, which, through oxidation reactions, activates the xenobiotic with the introduction of polar functional groups. The substratum is thus activated and allows the attack on the part of the conjugating enzymes (Phase II). MFOs are inducible by the xenobiotics present in the environment; therefore, they represent a qualitative or semi-quantitative signal of their presence. Their induction is specific to a given contaminant. Polycyclic aromatic hydrocarbons (PAHs) induce only one enzyme class (cytochrome P-4501A1), while the organochlorine insecticides (DDT, aldrin, etc.) induce a different enzyme family (P-4502B). Such high specificity of the MFOs allows researchers to obtain exact information regarding the cause of the polluting event. MFO induction has shown excellent results in monitoring aquatic ecosystems exposed to the waste products of paper manufacture. The pulp mills produce complex halogenated mixtures that induce the MFO system which can be measured *via* several tests [i.e. the induction of 7-ethoxyresorufin O-deethylase (EROD)] [10]. Liver EROD activity in European eel (*Anguilla anguilla*) is, for example, a suitable biomarker to benzo[*a*]pyrene (B[*a*]P) and *β*-naphthoflavone (BNF) exposure. A dose–response correlation to B[*a*]P and BNF was obtained (laboratory study) at the concentration range of 0.1–10 ppm [14]. Results of this study suggested that the eel could have a high ability for monitoring PAHs. The measurement of EROD activity in fish is a well-known *in vivo* biomarker of exposure to several HHCs and PAHs and some structurally related compounds [10].

Another specific biomarker is oxidative *stress*. Contaminants can increase the level of oxidative *stress* in aquatic organisms. Sources of potential production of free radicals induced by pollutants are redox reactions, which concern transition metals and organic free radicals. The organisms respond to the polluting event by developing antioxidant defence mechanisms capable of removing the free radicals.

Several molecules with a low molecular weight have this function, such as for instance, beta-carotene and vitamins A, E and C and other specific enzymes such as superoxide dismutase (SOD), catalases (CAT), glutathione peroxidase (GPx), glutathione S-transferase (GST), glutathione reductase (GR), nicotinamide adenine dinucleotide phosphate-oxidase (NADPH), DT-diaphorase (DT-d), glutathione (GSH) and lipid peroxidation.

Wu and Lam [15] have estimated the activities of glucose-6-phosphate dehydrogenase (G6PHD) and lactated dehydrogenase (LDH) in the adductor muscles of *Perna viridis* in a well-oxygenated site in Hong Kong. They demonstrated a significant negative correlation between these respiratory enzymes and ambient oxygen levels for *P. viridis*. The increase in LDH and G6PDH activities in response to hypoxia confirmed the hypothesis of these enzymes as good biomarkers for the monitoring of dissolved oxygen (DO) levels in marine waters.

Many studies [16, 17] conducted on *P. viridis* indicated that the majority of antioxidant parameters were induced by increasing tissue concentrations of PAHs (SOD, DT-d and lipid peroxidation did not show an exposure-concentration response, [16, 18]). Also biomarkers in aquatic plants have been well reviewed [19].

CAT activity increased significantly in *Posidonia oceanica* following a 48 h exposure to low concentrations of $HgCl_2$ (0.01 $\mu g\,L^{-1}$ and 0.1 $\mu g\,L^{-1}$) as compared to controls [20]. Also the oxidative stress present in freshwater species grown at high salt concentrations seems to induce CAT, SOD and peroxidase activity [21].

The above-mentioned mechanisms (cytochrome P450 and oxidative *stress*) concerning the metabolism of contaminants show a huge production of highly reactive intermediate reaction products. These metabolites of the contaminants can cause damages to the DNA. When these damages are not correctly repaired through the intervention of specific nuclear enzymes, there will be an incorrect gene expression (for instance, the oncogene activation). This process is characteristic of the initial phase of the chemical processes that trigger off the transformation of a normal cell into a tumour cell.

As already said, MTs (Fig. 2) were identified in approximately 50 species of aquatic vertebrates, particularly molluscs and crustaceans [22]. The function of

METALLOTHIONEIN

Molecular weight :	6100+ metal
Molecular shape :	non-globular
Chemical composition	
Total of amino acids:	61 residues/mol
cystein sulphur :	20mol/mol
thiolate sulphur :	20mol/mol
thiol sulphur :	none
disulfide sulphur :	none
aromatic amino acids :	none
Sum of metals (Zn + Cd) :	7mol/mol
Thiolate sulphur / sum of metals :	ca. 3

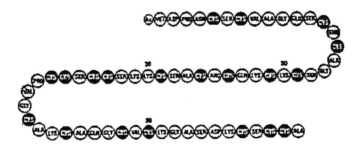

Figure 2: Metallothionein characteristics.

Table 1: Biomarker of trace metal exposure (modified from Depledge *et al.* [24]).

Biomarker	Contaminants initiating response
Metallothioneins	Cu, Zn, Cd, Co, Ni, Bi, Hg, Ag
Stress proteins	Cu and TBT
Glutathionic trasferases	Cd
Lipid peroxidation	Cd
Haem and porphyrins	Pb, As, Hg

MTs concerns both the preservation of homoeostasis and the processes of metal detoxification. MTs have a high cysteine content (30%), a low molecular weight, are stable to heat and have a strong affinity for some metals such as Ag, Cu, Cd, Hg and Zn. Generally, the degree of induction of MTs varies according to species and tissue. MTs induction in molluscs has been widely studied and can be correlated with Cd and Cu concentrations, as demonstrated by some studies [23]. MTs are detectable through several analytical methods such as differential pulse polarography, spectrophotometry, radioimmunological analysis.

Metallothionein induction has been widely demonstrated in studies carried out on microcosms, in sites exposed to metal contamination. However, this MT induction is less evident in studies performed *in situ*. It plays an important role both in the regulation of essential metals and in detoxification processes. The knowledge we have on the subject at present does not allow us to utilize MTs as biomarkers. Specific studies on aquatic organisms are particularly necessary. Table 1 reports biomarkers of trace metal exposure [24].

Esterases consist of two classes. Esterases A are responsible for the detoxification of organophosphorous insecticides while esterases B are inhibited in the presence of this kind of insecticides. The inhibition of brain esterases (AchE) and blood esterases (BchE and CbE) represents a specific indicator of OPs and CBs insecticide exposure.

The cholinesterase-inhibiting activity has been studied to diagnose for instance OPs pesticide poisoning in birds [13]. The Antarctic bivalve *Adamussium colbecki* showed the presence of organophosphorous-sensitive cholinesterases (ChE) and CbE activities in their gills [25].

In general, heavy metals and some pesticides (OPs and CBs) are considered to inhibit AchE, an essential enzyme in the signal transmission in the nervous system in the tissues of fish and mussels. Hence, the development of neurotoxicity biomarkers, which are measured by means of immunochemical diagnosis or electrophoretic techniques and are employed as general biomarkers to verify the *stress* level of an organism.

Stress proteins are produced at cellular level following a stressful event, as for instance the presence of contaminants. At the same time, the synthesis of these proteins provokes the inhibition of normal protein synthesis [26, 27]. These proteins are called heat shock proteins (HSPs). HSP70 is among the most commonly studied biomarkers of pollution, particularly in animals [28].

Lewis *et al.* (2001) [27] studied HSP70 expression in the seaweed *Enteromorpha intestinalis*. The alga was exposed to copper under different conditions of nutrient availability and the triazine herbicide Irgarol 1051. The HSP response was clearly affected by nutrient limiting conditions, and copper induced an increase in HSP70 only under nutrient replete conditions. Also, these authors [27] have demonstrated that HSP70 is only induced by stressors that are strongly proteotoxic. Results showed that HSP70 was not useful as a biomarker for copper and triazine.

Some laboratory studies have demonstrated that tributyltin (TBT) induces stress protein synthesis in *Mytilus edulis* [29] but there are some difficulties in extrapolating from the laboratory to the field. Lundebye *et al.* [30] conducted a research with the aim to evaluate the potential usefulness of stress proteins (stress-70 and stress-60 (chaperonin)) as a biomarker of cell injury in *M. edulis* exposed to varying degrees of organotin contamination *in situ*. The relative stress proteins measured did not reflect the organotin concentrations (TBT and dibutyltin (DBT)) measured in the gill tissue samples collected in five sites in the Fyn Island, Denmark, while the same correlation turned out to be positive between mussel tissues and sediments.

It is therefore right to stress that not always the results yielded by laboratory tests agree with the results of tests *in situ*, since in a natural environment the influence of interfering factors is much more common than in a laboratory situation where experimental conditions are established beforehand.

2.3   Metabolism products

Other biomarkers were identified in the products of metabolism such as for instance porphyrins, whose metabolism (especially in the liver) is altered by contaminants such as PCB, HCB and Pb. The increase in the concentrations of the intermediate products in the metabolism of porphyrins can be correlated with the presence of some classes of contaminants. In this instance, the biomarker is represented by the high hepatic concentrations of a number of intermediate metabolics such as protoporphyrins, uroporphyrins and coproporphyrins.

Traditionally, the analysis of the porphyrins is carried out on the liver of the animals (destructive biomarkers) residing in contaminated sites. The current trend is to use the porphyrins analysis in excreta and other biological fluids (i.e. blood) as a non-destructive biomarker. Casini *et al.* [31] and Fossi *et al.* [32] have established that the use of biological materials were useful for porphyrin determinations as a non-destructive biomarker for the hazard assessment of birds and sea lions contaminated with PCBs and other contaminants.

2.4   Hysto-cytopathological biomarkers

Many polluting compounds cause histopathological alterations in target organs, particularly in the liver. Generally, these changes affect the structure of cells, tissues, organs or individual organisms that are easy to measure. Here are a few examples of such biomarkers.

In this context human placenta has been proposed as a "dual" biomarker (foetal and maternal) for monitoring toxic trace elements exposure [33]. Placental Pb seems to be a useful tool as indicator of Pb exposure (long and short term) to assess maternal health during pregnancy [34]. Dennis and Fehr [35] demonstrated that there is a positive correlation between maternal and foetal Hg blood; particularly on women consuming high quantities of fish. Thus, monitoring the maternal and foetal blood for Hg is an effective method to control Hg exposure. For monitoring MeHg human hair is a better specimen than placenta [36].

Imposex is the development of male sexual characteristics in females (a penis and a vas deferens) [37]. The cause of imposex is attributed to exposure to TBT used in antifouling paints. In female of the neogastropod *Nucella lapillus* the development of imposex is induced by TBT concentrations of less than 1 ng Sn L^{-1} [37]. This specific sensitive response in neogastropods has been used as indicator of TBT contamination.

The vas deferens sequence index (VDSI) is one of the most important indices for biomonitoring TBT pollution [38]. VDSI quantifies the degree of imposex in the animal through the sequence of the penis and vas deferens formation (this scale is composed of 7 assessment level) [39].

The relative penis length index (RPLI) is an index that quantifies the degree of imposex in the population and is obtained from the equation: (Mean length of female penis)/(Mean length of male penis) $\times$ 100. This index is better applied in low contaminated areas.

The relative penis size index (RPSI) quantifies the degree of imposex in the population by the equation: (Mean length of female penis)^3/(Mean length of male penis)^3 $\times$ 100. This cubical index is better applied in highly contaminated areas, when the length of the female penis approaches the length of the male penis [40].

Mediterranean gastropods (particularly *Hexaplex trunculus*) and many other gastropod species (i.e. tropical muricids such as *Thais distinguenda* [41]; *Hinia reticulata* [42] showed also high sensitivity to imposex and were proposed as good bioindicators of TBT contamination [43]. The northwestern coasts of Sicily (Italy) have shown a widespread occurrence of imposex, reaching in most cases a 100% frequency [44].

TBT is not the sole cause of imposex. It can be caused also by other contaminants such as triphenyltin or nonylphenol [45]. However, Bryan *et al.* [46] underline that the measurement of imposex levels is a product of exposure over a lifetime. Thus, it cannot reflect the present pollution levels because imposex in gastropods is presumed to be irreversible.

Many contaminants can destabilize lysosomal membranes and membrane damage is often proportional to the magnitude of stress [47]. Lysosomes in the digestive gland of mussels constitute main sites of toxic metal and organic pollutant's accumulation and detoxification [47]. These cytological responses can be employed as useful non-specific biomarkers.

Lysosomal membrane stability (LMS) is the main lysosomal response to a wide range of pollutants and it can be considered a reliable biomarker of general stress in biomonitoring studies [48]. The LMS test is based on the cytochemical detection of the lysosomal enzyme N-acetyl-β-hexosaminidase, and the time of acid labilization treatment required producing lysosomal maximum staining intensity. The test is carried out on digestive gland cryosections (minor values of the technique correspond to more polluted marine areas, while higher values indicate less polluted areas).

The Neutral Red Retention (NRR) assay is used to assess lysosome responses in living mussel haemocytes [49]. This is a rapid and sensitive technique for determining the LMS against environmental toxic agents. The NRR measures the retention of neutral red, a weak base dye, within the lysosomal compartment. The test is carried out on isolated digestive cells and haemocytes; lower values correspond to mussels from polluted areas, while higher values to mussels from clean areas. Nicholson [50] has established that NRR assay in *Perna viridis* seems to be robust under laboratory conditions, and should be an effective non-destructive means of assessing metal pollution in subtropical marine waters. Recent studies [51] indicate some correlation between total PCB and total PAH body burdens and LMS for *P. viridis* in samples collected at coastal sites of Hong Kong.

Other applied biomarkers are related to morphometric evaluation of the number and size of the lysosomes, the neutral lipids and lipofuscins in mussels.

Histopathological changes, i.e. visible abnormalities in fish, have been used as indicators of adverse biological effects of toxic exposure; they reveal a state of disease. These biomarkers can be divided in morphological changes (external visible) and pathological (internal disease) [10].

In fish, externally visible changes can include: (1) find erosion; (2) skeletal malformation; (3) epidermal papilloma/hyperplasia; and (4) operculum abnormalities [10, 52].

Dab (*Limanda limanda*) is the most frequent flatfish of the North Sea and it is most frequently afflicted with externally visible diseases. Dab is a bottom dwelling flatfish which despite its massive occurrence in wide areas of the North Sea and Irish Sea, etc., is not of importance to fisheries [52].

An important disease present on a variety of different marine and fresh water fish species is *Lymhpocystis*. This is the most frequent external disease of dab. It is caused by a virus that provokes hyperthorphy of connective tissue cells and then the formation of nodules [52]. The nodules can be situated anywhere on the body surface.

Epidermal papilloma/hyperplasia are benign tumours affecting stroma and other tissues of the epidermis with unknown aetiology. Another disease is the ulceration that is probably caused by bacteria and can be influenced by poor nutrition, salinity fluctuations and pollution [52].

Internal disease indicators refer to histological biomarkers. They provide, together with fish diseases, a powerful tool to characterise the biological end points of toxicant and carcinogen exposure [53, 54].

Of all the organs, the liver is the most extensively studied because it is the target of xenobiotic metabolism and detoxification mechanisms.

As reported by Lam and Wu [10] typical symptoms of pathological lesions observed in liver of animals collected from polluted sites are:

1. Neoplasms;
2. Putatively pre-neoplastic lesions;
3. Non-neoplastic proliferative lesions;
4. Specific (unique) degenerative or necrotic lesions;
5. Non-specific necrotic lesions unassociated with visible infectious agents;
6. Hydropic vacuolation of hepatocytes or biliary epithelial cells.

Many laboratory and field studies have shown an association between specific chemical exposure and tumour induction in fish [55–57].

Reynolds *et al.* [58] studied the response of several biomarkers and histopathological indicators of PAHs exposure in the flounder (*Platichthys flesus* L.). Flounder were fed daily with food spiked with a mixture of four PAHs at an environmentally relevant range of concentrations for either one or 6 months. EROD activity was elevated following 1 month exposure to PAH concentrations up to 50 mg kg^{-1} in food. Bile metabolite concentrations (measured as 1-OH pyrene equivalents) were found to increase with PAH concentration, up to 500 mg kg^{-1} PAH. After the 6 months PAH exposure, EROD levels were not elevated, bile metabolites showed a similar dose dependant relationship as in the 1 month experiment. No significant histopathological changes were observed.

The authors of this study support the utility of European flounder as a sentinel species for the measurement of biomarker responses to PAHs. Measures such as EROD activity and bile metabolite concentration may provide a useful tool when considering short-term exposures to chemicals such as PAH. In cases of chronic or historic contaminant exposure, the measurement of other biomarkers, such as the presence of DNA adducts, indicates the initiation of carcinogenesis. On the contrary, histopathology provides a robust endpoint of carcinogen exposure.

The authors [58] concluded that a multi-biomarker approach to aquatic pollution monitoring allows the best assessment of whole animal response to a range of anthropogenic contaminants.

Biomarkers are also used for the assessment of the potential negative effects of chemicals in soil ecosystems [59].

Experimental exposure of the woodlouse *Porcellio laevis* (Isopoda) to cadmium sulphate resulted in changes in the structure of the hepatopancreas, that is the target organ for metal accumulation [60]. In this study the quantification of changes in hepatopancreas structure indicated that there was a dose-related change in the histological sections of the hepatopancreas of *P. laevis* exposed for 6 weeks to 20, 80 and 160 mg kg^{-1} cadmium sulphate-exposure groups. These histological biomarkers are relatively easy to detect, but they may not be very useful for regular monitoring. They can serve as general biomarkers of metal exposure in Isopoda species.

Gill histopathology has also been shown to be responsive to a wide range of pollutants. Gills are continually exposed to contaminants and are the first-line defence against water borne insults. They perform many critical physiological functions such as gaseous exchange, osmoregulation, excretion, etc. [10].

For instance, the amoebic gill disease (AGD) is a non-destructive and reliable method (and possible biomarker) for rapid and broad-scale disease management of farmed Atlantic salmon, *Salmo salar* L. [61]. However, this technique has been considered questionable due to the difficulty to establish a precise correlation between clinical signs and histological lesions observed in a commercial setting.

In general, histopathological biomarkers appear to be useful for environmental monitoring. However, they would be used together with other kinds of biomarkers in environmental programmes.

2.5   Immunological biomarkers

Many acquatic species present multifaceted responses of their immune system to immune moderators, particularly to xenobiotics. Several advantages can be indicated for this kind of biomarker. For example, responses occur when chemical concentration is low [62]. Also, they provide evidence linking a toxicant to disease outbreak in fish and blood samples can be taken over a long period of time in order to allow the assessment of a toxin [62]. On the other hand, there are some limitations: data cannot be extrapolated between species; the immune response sometimes has a low specificity and it is difficult to achieve reliable results; xenobiotics must already be known, etc. [62].

Many immunotoxicity assays have been developed in invertebrates [63, 64]. Ville *et al.* [65, 66] used a biomarker of humoral and cellular immunodefence responses of earthworms, using an assay of serine protease activity and a serine protease inhibitor to test the effects of PCBs, Carbaryl (carbamate) and 2,4-dichloro phenoxyacetic acid (2,4 D). The contact tests assay showed that carbaryl is characterized by a low LC 50 value of 3.4 μg cm^{-2}, compared to 18 μg cm^{-2} for 2,4 D and 32 μg cm^{-2} for PCBs. Responses mediated by free proteins, detected through *in vitro* assays for lysozyme, haemolysis, protease and serpin activities were modified in earthworms. Also, non-specific immunity was studied, like phagocytosis, that is decreased dramatically in nearly all earthworms exposed to soil contaminants.

The hierarchical approach can be used in order to classify immunological indicators. The first class of assays considers the apparatus for immunity, i.e. the cells and tissues involved, the second investigates the mechanisms involved and the third estimates the efficacy of the response in terms of disease susceptibility. Thus, immunological indicators can be classified into a three-tier regime.

Tier 1 tests involve a general screening of the immune system. These studies are related to gross morphology or cellular conditions of the organs of the immune system such as the weight and size measurements of the spleen, haematocrit, leukocrit, wound healing, macrophage phagocytosis, lysozyme activity, agglutination assay, graft rejection, percentage of circulating antibodies activated by lysozyme [62]. These assays are inexpensive and easy to conduct but they are influenced by temperature, handling and crowding of the organisms [62].

Tier 2 tests regards a more comprehensive evaluation of the immune system. A variety of assays can be included in this category such as immune cell quantification, native immunoglobin quantification, surface markers, phagocytic index (that are the cells involved in the initial immunal response) and plaque-forming cell assay. Also the mitogenic response can be measured. It can be defined as the index of factors stimulating lymphocytes B or T proliferation [62].

Tier 3 tests involve the resistance of the host to a stressor. They involve host resistance challenge studies to syngeneic tumours, bacterial, viral and parasitic infectivity models [64, 67].

The first two classes of assay have been used quite widely for marine bivalves while the third class of assay were recently included in studies on invertebrate immunotoxicology [63].

Many studies on environmental modulation of the immune function in marine bivalves have focused on heavy metals; however, the effects of organic compounds have also been addressed in both laboratory and field studies [63].

2.6   Physiological biomarkers

There are not many studies on physiological biomarkers because of the difficulty in the continuous measurements which must allow for the speed of the various processes involved. Many studies in this field concern the breathing activity, the cardiac cycle, osmoregulation, growth and energy metabolism as well as the behaviour of some species (fish, molluscs, crustaceans, etc.) exposed, for instance, to different metal concentrations. Also the measurement of the blood parameters (protein concentrations, coagulation time) can be used to monitor the condition of animals in an aquatic environment.

The measurement of the Scope for Growth (SFG) [68] is one of the possible indexes of physiological integrity. This index is an integrative measurement of the energetic status of an organism and it has been largely employed in ecotoxicology studies. Monitoring growth is one of the most sensitive ways to measure *stress* in an organism. It represents the integration of physiological responses to all possible man-induced impacts. SFG is expressed as the balance between the intake processes (nutrition and digestion) and energy output (metabolism and excretion). These physiological responses can be converted in measurements of energy flow ($J g^{-1} h^{-1}$).

SFG depends on a series of physiological or biochemical factors, such as changes in the speed of food consumption for each species and food availability, changes in the efficiency of the digestion processes and nutrient absorption, as well as an increase or a decrease in the breathing speed of the organisms analyzed. For instance, Widdows *et al.* [68] detected huge levels of SFG ($16 J g^{-1} h^{-1}$) in mussels (*Mytilus galloprovincialis*) collected in an external site of the Venice Lagoon that was relatively little contaminated if compared to sites inside the lagoon which showed significantly reduced values of SFG (2 to $4 J g^{-1} h^{-1}$) and showing therefore severe pollution-induced stress.

Also the DNA/RNA ratio can be measured for growth measurements. The rate of protein synthesis is related to RNA and consequently to DNA concentrations.

Given that DNA concentrations tends to remain constant among cells, a higher ratio of RNA relative to DNA or protein content can reflect a better growth of an organism [10].

Wo *et al.* [69] demonstrated that exposure to sublethal concentrations of cadmium (0.93–2.19 mg L^{-1}) for 8 days caused a significant reduction in the RNA/DNA ratio in a gastropod *Nassarius festivus*. Besides, measurements based on SFG revealed significant deviation from the control when the *N. festivus* were exposed to 0.16 mg L^{-1} cadmium or higher for 8 days.

The authors also conducted growth measurements based on shell size and body weight that is a simple and inexpensive method. Results of this experiment show evidence of effects of cadmium exposure on the rate of increase in shell size and body weight when experiments were extended to 16 days [69]. Results of this study suggest that SFG is the most sensitive biomarker, followed by the RNA/DNA ratio, and then the measurements based on shell size and body weight [69].

Growth rate is a fundamental component of the physiology of organisms and it is therefore an important index of the effect of environmental contaminants.

The determination of the energy available for the growth on the basis of the physiological analysis of the energy balance provides us with an immediate evaluation of the energy condition of the animal and of the different components influencing the growth rate (breathing, energy load, etc.).

For instance, many ecotoxicology studies are based on the registration of the variations in the physiological responses caused by the toxicity of the polluting substances against autotrophic organisms: adverse effects on vital functions such as growth or photosynthesis are an index of the integral toxicity of the aqueous medium.

Eukaryotic or procaryotic microalgae, crustaceans, amphibians, etc., have been employed so as to obtain biomarkers capable of giving real-time information on drastic changes in the quality of the waters, following the introduction of pollutants in the aquatic ecosystem.

By way of example, the crustacean *Daphnia magna* is often utilized since it is highly sensitive to a wide number of aquatic contaminants [70] just like mussels [63].

The impact of the contaminants on the reproductive success of a population is very important since it can produce dangerous effects. Generally, in these instances, the contaminants affect the early stages of the development of an organism. For instance, the pollution caused by organostannic compounds in France (Arcachon bay) a few years ago, in an oyster production facility, disrupted the larval colonies for many years, causing phenomena of malformation [71, 72].

In general, there is an evidence of the general assumption that molecular/biochemical responses usually precede physiological/behavioural responses in ecotoxicological studies.

2.7   Behavioural biomarkers

Changes in the behaviour of organisms under different ecological conditions can be used to indicate their responses to environmental stressors. This kind of biomarkers cannot be used as an early warnings system of ecological stress.

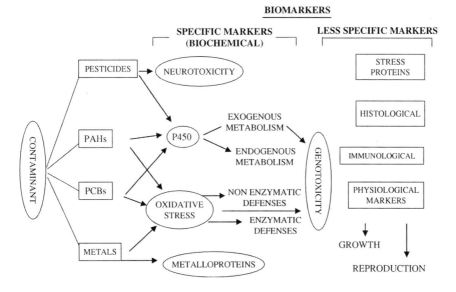

Figure 3: Interactions between contaminants and biomarkers (modified from Narbonne and Michel, 1993 [1]).

With the help of a video camera connected to a special software (a special experimental aquarium) Depledge and Aagaard [24] studied the possibility to observe the motor reactions of a number of animals (terrestrial invertebrates and crustaceans) exposed to a number of contaminants (trace metals). This approach makes it possible to highlight the sensitivity and the specificity of the behavioural change. Moreover, it allows to observe the possible response times to the action of a contaminant at a high hierarchical level.

In this kind of studies it is possible to conduct several different measurements such as sensory responses (i.e. phototaxis, chemotaxis, larval settlement, temperature preferences, tactile inhibition), rhythmic activities, motor activities, interindividual responses (migration, aggression and predation vulnerability, etc.) [10]. Other studies show parameters such as the distance covered by an animal, mean speed, motor inactivity time, movement frequency and the relationships between these parameters.

The main interactions between contaminants and biomarkers are reported in Fig. 3.

3 Conclusions

The development of biomarkers (and bioindicators) in the environmental field for the preventive identification and the prevention of harmful effects is highly needed.

It is therefore extremely important to use biomarkers towards the prevention of events that are harmful both to man and ecosystems. Biomarkers and bioindicators could be used also as potential tools for environmental quality assessement during judicial proceedings.

One biomarker alone might not be sufficient to monitor the environmental quality. In general, it is necessary to have a battery of biomarkers allowing the evaluation of the physiological integrity of all species present in a given ecosystem. Batteries of biomarkers should be capable to assess the toxic effects of the highest possible number of environmental contaminants [73].

Generally, biomarkers do not show one single dose–response relationship but a number of dose–response relationships that are the result of all possible interactions of all contaminants present in the ecosystem under study. So, battery of biomarkers allows to determine at what level of structural complexity a given response to a toxic compound in a given ecosystem occurs.

In brief, the multiple response concept shows that through biomarkers it is possible to determine the "health level" of the population studied. The change in the homoeostasis present in an ecosystem trigger off a series of compensatory mechanism both at the biochemical and the physiological level. These mechanisms tend to restore the initial homoeostatic situation. The more toxic concentrations or time of exposure increase the more the responses turn out to be insufficient with subsequent occurrence of other responses at higher levels all the way up to significant phenomena, such as bacterial infections, parasitism, carcinogenesis and finally the death of the organisms. Every response, at every level of complexity (biochemical, physiological, etc.) is considered a biomarker. Therefore, the use of batteries of biomarkers allows us to identify the level of a given organism, and, according to the results, the risk level for the whole population. Risk levels can be as follows [74]:

1. homeostasis
2. compensatory responses
3. bioremediation
4. disease

It is rather difficult to provide an experimental definition of these steps and in fact many scientific researches have been working towards trying to fill this gap. There are several interpretation models on this issue. Scientific research on biomarkers tries to identify those biomarkers that have a high ecological level, that is biomarkers that are linked with specific changes in the Darwinian fitness (i.e. growth alterations, reproductive insuccess, variation in the energetic balance, etc.).

Biomarkers are also very useful in the bioremediation or environmental restoration [74]. With respect to, this it is very important to adopt a non-destructive approach that does not involve the sacrifice of animals. These techniques allow to monitor the modification of some biochemical (and/or molecular) parameters in the different phases of environmental restoration without damaging the organism.

For this purpose, blood, faeces, milk, hairs, horns, eggs, feathers and hepatic or skin biopsy are employed in order to evaluate the health of a population.

In conclusion, current researches tend to identify, develop and verify biomarkers that are highly sensitive, inexpensive and easy to use. As reported above, other important biomarkers are those that have a close relationship with the Darwinian fitness and those capable of identifying the responses at the different levels of structural complexity. One more important development is the study of biomarkers in the field of environmental restoration within of monitoring programs.

References

[1] Narbonne J.-F., Michel X. (1993) Use of biomarkers in the assessment of contamination in marine ecosystems. Fundamental approach and applications, in: Selected techniques for monitoring biological effects of pollutants in marine organisms. UNEP-MAP Technical Reports Series, 71, 1–20.

[2] Santagostino A., Conte M., Fumagalli P., Galvani P., Zanolli L. (1997) Organismi e sistemi enzimatici indicatori di stress antropogenici nel comparto suolo-aria. Acqua & Aria, 6, 115–118.

[3] Conti M.E. (2002) Il monitoraggio biologico della qualità ambientale. ed. Seam, Rome, p. 185.

[4] Lam P.K.S., Gray J.S. (2003) The use of biomarkers in environmental monitoring programmes. Marine Pollution Bulletin, 46, 182–186.

[5] Depledge M.H. (1994) The rational basis for the use of biomarkers as ecotoxicological tools, in: Fossi M.C., Leonzio C. (Eds.), Nondestructive Biomarkers in Vertebrates. Lewis Publishers, CRC PressBoca Raton, FL, USA, pp. 272–295.

[6] Depledge M.H. (1996) Interpretation, relevance and extrapolations: Can we devise better ecotoxicological tools to assess toxic impacts? in: Tapp J.F., Warfe J.R., Hunt S.M. (Eds.), Toxic Impacts of Wastes on the Aquatic Environment. Royal Society of Chemistry, Cambridge, UK, pp. 104–115.

[7] Matsuba T., Keicho N., Higashimoto Y., Granleese S., Hogg J.C., Hayashi S., Bondy G.P. (1998) Identification of glucocorticoid and adenovirus E1A-regulated genes in lung epithelial cells by differential display. American Journal of Respiratory Cell and Molecular Biology 18, 2, 243–254.

[8] Timbrell J.A. (1998) Biomarkers in toxicology. Toxicology, 129, 1, 1–12.

[9] Liang P., Pardee A.B. (1992) Differential display of eukaryotic messenger RNA by means of the polymerase chain reaction. Science, 257(5072), 967–971.

[10] Lam P.K.S., Wu R.S.S. (2003) Use of biomarkers in Environmental Monitoring, STAP Workshop on: The use of bioindicators, biomarkers and analytical methods for the analysis of POPs in Developing Countries, 10–12 December 2003.

[11] Denslow N.D., Bowman C.J., Ferguson R.J., Lee H.S., Hemmer M.J., Folmar L.C. (2001) Induction of gene expression in sheepshead minnows

(*Cyprinodon variegatus*) treated with 17β-estradiol, diethylstilbestrol, or ethinylestradiol: the use of mRNA fingerprints as an indicator of gene regulation. General and Comparative Endocrinology, 121(3), 250–260.

[12] Shugart L.R., (1999) Structural damage to DNA in response to toxicant exposure, in: Forbes V.E. (Ed.), Genetics and Ecotoxicology. Taylor & Francis, U.S., pp. 151–168.

[13] Walker C.H. (1995) Biochemical biomarkers in ecotoxicology – some recent developments. Science of the Total Environment, 171, 189–195.

[14] Bonacci S., Corsi I., Chiea R., Regoli F., Focardi S. (2003) Induction of EROD activity in European eel (*Anguilla anguilla*) experimentally exposed to benzo[a] pyrene and β-naphthoflavone. Environment International, 29(4), 467–473.

[15] Wu R.S.S., Lam P.K.S. (1997) Glucose-6-phosphate dehydrogenase and lactate dehydrogenase in the green-lipped mussel (*Perna viridis*): possible biomarkers for hypoxia in the marine environment. Water Research, 31(11), 2797–2801.

[16] Cheung C.C.C., Zheng G.J., Li A.M.Y., Richardson B.J., Lam P.K.S. (2001) Relationships between tissue concentrations of polycyclic aromatic hydrocarbons and antioxidative responses of marine mussels, *Perna viridis*. Aquatic Toxicology, 52, 189–203.

[17] Lau P.S., Wong H.L. (2003) Effect of size, tissue parts and location on six biochemical markers in the green-lipped mussel, *Perna viridis*. Marine Pollution Bulletin, 46, 1563–72.

[18] Nicholson S., Lamb P.K.S. (2005) Pollution monitoring in Southeast Asia using biomarkers in the mytilid mussel *Perna viridis* (Mytilidae: Bivalvia). Environment International, 31, 121–132.

[19] Ferrat L., Pergent-Martini C., Romeo M. (2003) Assessment of the use of biomarkers in aquatic plants for the evaluation of environmental quality: application to seagrasses. Aquatic Toxicology, 65, 2, 187–204.

[20] Ferrat L., Roméo M., Gnassia-Barelli M., Pergent-Martini C. (2002) Effect of mercury on antioxidant mechanisms in the marine phanerogam Posidonia oceanica. Diseases of Aquatic Organisms, 50(2), 157–160.

[21] Rout N., Shaw P. (2001) Salt tolerance in aquatic macrophytes: possible involvement of the antioxidative enzymes. Plant Science, 160(3), 415–423.

[22] Cajaraville M.P., Bebianno M.J., Blasco J., Porte C., Sarasquete C., Viarengo A. (2000) The use of biomarkers to assess the impact of pollution in coastal environments of the Iberian Peninsula: a practical approach. Science of the Total Environment, 247, 295–311.

[23] Bebbiano M.J., Machado L.M. (1997) Concentrations of metals and metallothioneins in *Mytilus galloprovincialis* along the south coast of Portugal. Marine Pollution Bulletin, 34, 666–671.

[24] Depledge M.H., Aagaard A., Györkös P. (1995) Assessment of trace metal toxicity using molecular, physiological and behavioural biomarkers. Marine Pollution Bulletin, 31, 19–27.

[25] Bonacci S., Browne M.A., Dissanayake A., Hagger J.A., Corsi I. (2004) Esterase activities in the bivalve mollusc *Adamussium colbecki* as a biomarker for pollution monitoring in the Antarctic marine environment. Marine Pollution Bulletin, 49(5–6), 445–455.

[26] Mc Carthy J.F., Shugart L.R. (1990) Biomarkers of Environmental Contamination. Lewis, Boca Raton, FL.

[27] Lewis S., Donkin M.E., Depledge M.H. (2001) Hsp70 expression in Enteromorpha intestinalis (Chlorophyta) exposed to environmental stressors. Aquatic Toxicology, 51(3), 277–291.

[28] Lewis S., May S., Donkin M.E., Depledge M.H. (1998) The influence of copper and heatshock on the physiology and cellular stress response of Enteromorpha intestinalis. Marine Environmental Research, 46(1–5), 421–424.

[29] Steinert S.A., Pickwell G.V. (1993) Induction of HSP70 proteins in mussels by ingestion of tributyltin. Marine Environmental Research, 35(1–2), 89–93.

[30] Lundebye A.K., Langston W.J., Depledge M.H. (1997) Stress proteins and condition index as biomarkers of tributyltin exposure and effect in mussels. Ecotoxicology, 6(3), 127–136.

[31] Casini S., Fossi M.C., Cavallaro K., Marsili L., Lorenzani J. (2002) The use of porphyrins as a non-destructive biomarker of exposure to contaminants in two sea lion populations. Marine Environmental Research, 54, 769–773.

[32] Fossi M.C., Casini S., Marsili L. (1996) Porphyrins in excreta: a nondestructive biomarker for the hazard assessment of birds contaminated with PCBs. Chemosphere, 33(1), 29–42.

[33] Iyengar G.V., Rapp A. (2001) Human placenta as a dual biomarker for monitoring fetal and maternal environment with special reference to potentially toxic trace elements. Part 3: toxic trace elements in placenta and placenta as a biomarker for these elements. Science of the Total Environment, 280(1–3), 221–238.

[34] Diaz-Barriga F., Carrizeles L., Calderon J., Batres L., Yanez L., Tabor M.W., Castelo J. (1995) Measurement of placental levels of arsenic, lead and cadmium as a biomarker of exposure to mixtures, biomonitors and biomarkers as indicators of environmental change. Plenum Press, New York.

[35] Dennis C.A.R., Fehr F. (1975) The relationship between mercury levels in maternal and cord blood. Science of the Total Environment, 3(3), 275–277.

[36] Carrier G., Bouchard M., Brunet R.C., Caza M. (2001) A toxicokinetic model for predicting the tissue distribution and elimination of organic and inorganic mercury following exposure to methyl mercury in animals and humans. II. Application and Validation of the Model in Humans Toxicology and Applied Pharmacology, 171(1), 50–60.

[37] Gibbs P.E., Bryan G.W. (1996) TBT-induced imposex in neogastropod snails: masculinization to mass extinction, in: De Mora S.J. (Ed.), Tributyltin: Case Study of an Environmental Contaminant. Cambridge University Press, Cambridge (UK), pp. 211–236.

[38] Gibbs P.E., Brian G.W., Pascoe P.L., Burt G.L. (1987) The use of the dog-whelk, *Nucella lapillus*, as an indicator of tributyltin TBT contamination. Journal of the Marine Biological Association of the United Kingdom, 67(3), 507–523.

[39] Terlizzi A., Geraci S., Gibbs P.E. (1999) TBT-induced imposex in the Neogastropod Hexaplex trunculus in Italian coastal waters: morphological aspects and ecological implications. Italian Journal of Zoology, 66(2), 141–146.

[40] De Castro I.B., Meirelles C.A.O., Matthews-Cascon H., Fernandez M.A. (2004) Thais (Stramonita) Rustica (Lamarck, 1822) (Mollusca: Gastropoda: Thaididae), A Potential Bioindicator of Contamination by Organotin Northeast Brazil. Brazilian Journal of Oceanography, 52(2), 135–139.

[41] Bech M., Strand J., Jacobsen J.A. (2002) Development of imposex and accumulation of butyltin in the tropical muricid Thais distinguenda transplanted to a TBT contaminated site. Environmental Pollution, 119(2), 253–260.

[42] Santos M.M., Vieira N., Reis-Henriques M.A., Santos M.A., Gomez-Ariza J.L., Giraldez I., Ten Hallers-Tjabbes C.C. (2004) Imposex and butyltin contamination off the Oporto Coast (NW Portugal): a possible effect of the discharge of dredged material. Environment International, 30(6), 793–798.

[43] Axiak V., Vella A.J., Micaleff D., Chircop P. (1995) Imposex in *Hexaplex trunculus* (Gastropoda: Muricidae): first result from biomonitoring of TBT contamination in the Mediterranean. Marine Biology, 121(4), 685–692.

[44] Chiavarini S., Massanisso P., Nicolai P., Nobili C., Morabito R. (2003) Butyltins concentration levels and imposex occurrence in snails from the Sicilian coasts (Italy). Chemosphere, 50(3), 311–319.

[45] Birchenough A.C., Barnes N., Evans S.M., Hinz H., Krönke I., Moss C. (2002) A review and assessment of tributyltin contamination in the North Sea, based on surveys of butyltin tissue burdens and imposex/intersex in four species of neogastropods. Marine Pollution Bulletin, 44(6), 534–543.

[46] Bryan G.W., Gibbs P.E., Burt G.R., Hummerstone L.G. (1987) The effect of tributyltin (TBT) accumulation on adult dog-whelks, Nucella lapillus: long-term field and laboratory experiments. Journal of Marine Biological Association United Kingdom, 67(3), 525–544.

[47] Moore M.N. (1985) Cellular responses to pollutants. Marine Pollution Bulletin, 16(4), 134–139.

[48] Domouhtsidou G.P., Dailianis S., Kaloyianni M., Dimitriadis V.K. (2004) Lysosomal membrane stability and metallothionein content in Mytilus galloprovincialis (L.), as biomarkers: combination with trace metal concentrations. Marine Pollution Bulletin, 48(5–6), 572–586.

[49] Lowe D.M., Fossato V.U., Depledge M.H. (1995) Contaminant-induced lysosomal membrane damage in blood cells of mussels *Mytilus galloprovincialis* from the Venice Lagoon: an *in vitro* study. Marine Ecology Progress Series, 129(1–3), 189–196.

[50] Nicholson S. (2003) Lysosomal membrane stability, phagocytosis and tolerance to emersion in the mussel Perna viridis (Bivalvia: Mytilidae) following exposure to acute, sublethal, copper. Chemosphere, 52, 1147–1151.

[51] HKEPD. (2002) Development of a Biological Indicator System for Monitoring Marine Pollution. Agreement No. CE2/2001 (EP). Hong Kong Government. Hong Kong Special Administrative Region.

[52] ICES (1996) Common diseases and parasites of fish in the North Atlantic: training guide for identification. ICES Cooperative Research Report, 19, 1–27.

[53] Schwaiger J., Wanke R., Adam S., Pawert M., Honnen W., Triebskorn R. (1997) The use of histopathological indicators to evaluate contaminant-related stress in fish. Journal of Aquatic Ecosystem Stress and Recovery, 6(1), 75–86.

[54] Stentiford G.D., Longshaw M., Lyons B.P., Jones G., Green M., Feist S.W. (2003) Histopathological biomarkers in estuarine fish species for the assessment of biological effects of contaminants. Marine Environmental Research, 55(2), 137–159.

[55] Malins D.C., McCain B.B., Myers M.S., Roubal W.T., Schwiewe M.H., Landahl J.T., Chan S.-L. (1987) Field and laboratory studies of the etiology of liver neoplasms in marine fish from Puget Sound. Environ Health Perspect, 71, 5–16.

[56] Krause M.K., Rhodes L.D., Vanbeneden R.J. (1997) Cloning of the p53 tumor suppressor gene from the Japanese medaka (Oryzias latipes) and evaluation of mutational hotspots in MNNG-exposed fish. Gene, 189(1), 101–106.

[57] Jenkins J.A. (2004) Fish Bioindicators of Ecosystem Condition at the Calcasieu Estuary. Louisiana National Wetlands Research Center, USGS.

[58] Reynolds W.J., Feist S.W., Jones G.J., Lyons B.P., Sheahan D.A., Stentiford G.D. (2003) Comparison of biomarker and pathological responses in flounder (Platichthys flesus L.) induced by ingested polycyclic aromatic hydrocarbon (PAH) contamination. Chemosphere, 52(7), 1135–1145.

[59] Kammenga J.E., Dallinger R., Donker M.H., Kohler H.R., Simonsen V., Triebskorn R., Weeks J.M. (2000) Biomarkers in terrestrial invertebrates for ecotoxicological soil risk assessment. Reviews of Environmental Contamination and Toxicology, 164, 93–147.

[60] Odendaal J.P., Reinecke A.J. (2003) Quantifying histopathological alterations in the hepatopancreas of the woodlouse Porcellio laevis (Isopoda) as a biomarker of cadmium exposure. Ecotoxicology and Environmental Safety, 56, 319–325.

[61] Adams M.B., Ellard K., Nowak B.F. (2004) Gross pathology and its relationship with histopathology of amoebic gill disease (AGD) in farmed Atlantic salmon, Salmo salar L. Journal of Fish Diseases, 27(3), 151.

[62] Unep/Map (2004) Guidelines for the development of ecological status and stress reduction indicators for the mediterranean region. MAP Technical Reports Series, 154.

[63] Livingstone D.R., Chipman J.K., Lowe D.M., Minier C., Mitchelmore C.L., Moore M.N., Peters L.D., Pipe R.K. (2000) Development of biomarkers to detect the effects of organic pollution on aquatic invertebrates: recent

molecular, genotoxic, cellular and immunological studies on the common mussel (*Mytilus edulis* L.) and other mytilids, in: Conti M.E., Botrè F. (Eds.), The Control of Marine Pollution: Current Status and Future Trends. International Journal of Environment and Pollution, 13(1), 56–91.

[64] Galloway T.S., Depledge M.H. (2001) Immunotoxicity in invertebrates: measurement and ecotoxicological relevance. Ecotoxicology, 10(1), 5–23.

[65] Ville P., Roch P., Cooper E.L., Narbonne J.F. (1997) Immuno-modulator effects of PCBs, carbaryl and 2,4 D in the earthworm *Eisenia fetida andrei*. Developmental & Comparative Immunology, 21(2), 118.

[66] Ville P., Roch P., Cooper E.L., Narbonne J.F. (1997) Immuno-modulator effects of carbaryl and 2,4 D in the earthworm *Eisenia fetida andrei*. Archives of Environmental Contamination and Toxicology, 32(3), 291–297.

[67] Pipe R.K., Coles J.A. (1995) Environmental contaminants influencing immune function in marine bivalve molluscs. Fish & Shellfish Immunology, 5, 581–595.

[68] Widdows J., Nasci C., Fossato V.U. (1997) Effects of pollution on the scope for growth of mussels (*Mytilus galloprovincialis*) from the Venice Lagoon, Italy. Marine Environmental Research, 43(1), 69–79.

[69] Wo K.T., Lam P.K.S., Wu R.S.S. (1999) A comparison of growth biomarkers for assessing sublethal effects of cadmium on a marine gastropod, *Nassarius festivus*. Marine Pollution Bulletin, 39(1–12), 165–173.

[70] Bonfanti P., Colombo A., Urani C., Cantelli D., Camatini M. (1997) Biological markers for evaluation of water quality. Acqua Aria, 6, 77–82.

[71] Pagano M., Cipollaro M., Corsale G., Esposito A., Ragucci E., Giordano G.G., Trieff N.M. (1985) Comparative toxicities of chlorinated biphenyls on sea urchin egg fertilisation and embryogenesis. Marine Environmental Research, 17(2–4), 240–244.

[72] Tyurin A.N. (2000) Choice of biotests and bioindicators for evaluation of the quality of the marine environment, in: Conti M.E., Botrè F (Eds.), The Control of Marine Pollution: Current Status and Future Trends, International Journal of Environment and Pollution, 13(1–6), 45–55.

[73] Peakall D. (1992) Animal biomarker as pollution indicators. Chapman & Hall, London.

[74] Fossi M.C. (1998) Biomarker: strumenti diagnostici e prognostici di 'salute' ambientale, in: Ecotossicologia, Vighi M., Bacci E. (Eds.), UTET, Torino, pp. 60–73.

3 Biomonitoring of freshwater environment

M E Conti

1 Principles of the Water Framework Directive

The Water Framework Directive 2000/60 EC (WFD) [1] is an important opportunity to reorganize the basic integrated resource management. The key issue of the Directive is an extremely *wide-ranging* and *integrated* approach based on a *sustainable* use of the aquatic resource,

1. first of all, it introduces *quality* standards that will serve as points of reference for the quality assessment of bodies of water in the UE member countries. The objective is for the artificial and heavily polluted water bodies to regain a good ecological potential. In view of this, a 'maximum ecological potential' has to be established as a reference condition;
2. it therefore defines *quality* objectives to be met; quality classes have been ranked in a five-class-system ecological status (high, good, moderate, poor and bad);
3. it establishes criteria for the identification of significant water bodies and the assessment of their environmental quality status.

The environmental quality status is defined on the basis of the *chemical* and *ecological* status of the water body. For superficial water bodies, the assessment of the chemical status is carried out on the basis of the presence – beyond a given threshold value – of harmful substances. The ecological status is the expression of the various components of the aquatic ecosystem (physical and chemical nature of waters and sediments, characteristics of the water flow, structure of the river bed). The use of *bioindicators*, as effective means for the evaluation of human impact on the water body, is provided for in view of a classification of running water, within the framework of the definition of the ecological status.

Other relevant key issues of the WFD are the preservation and the restoration of the health of the aquatic ecosystems and not to preserve the different aspects of water utilization as previously reported in other water-related EU Directives.

WFD establishes that the assessment be based on biological monitoring approach, studying the aquatic communities, molluscs, algae, macrophytes, benthic macroinvertebrates and fish.

1.1 Water indicators

Many kinds of indicators are used for water quality assessment. The European Environmental Agency (EEA) has adopted the Drivers-Pressures-State-Impact-Responses (DPSIR) framework which represents a system approach of analysis of the relations between environment and human system.

According to the DPSIR framework, developments of a financial or social nature are the underlying factors or "Drivers" exerting "Pressures" on the environment the Status of which – availability of resources, level of biodiversity or quality of the waters – change accordingly. These factors may have an impact on human health as well as on ecosystems; these processes require a response able to modify the driving forces, reducing pressure and impacts [2]. Fig. 1 shows a generic DPSIR framework developed by EEA.

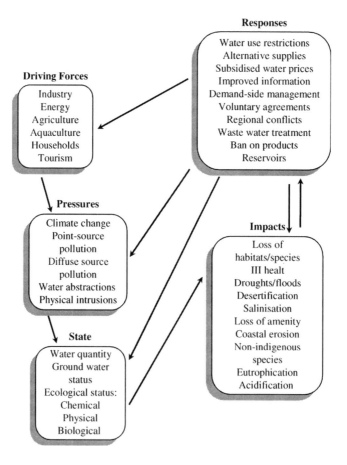

Figure 1: EEA DPSIR framework for water (modified from Nixon *et al.* [2]).

The DPSIR makes it possible to select water-specific issues in order to achieve relevant policy objectives. The EEA Water Assessment [2] has designed four relevant issues: ecological quality, nutrients and organic matter, hazardous substances and water quantity. Table 1 reports an abbreviated list of the EEA indicators for water.

The United Nations Environment Programme (UNEP) developed series of indicators related to environmental aspects linked to water, land, atmosphere and waste (Consultative Group on Sustainable Development [3]). Table 1 also shows UNEP indicators related to water.

2   Freshwater ecosystems

As mentioned above, the philosophy underlying the directive 2000/60/CE [1] aims at defining the ecological status of water bodies, focusing the attention on biological quality elements, also backed by hydro-morphological and chemico-physical elements.

This crucial conceptual turning point leads to considering rivers as ecosystems in their own right, that is "living systems" provided with structures and functions that need to be recovered and preserved in order to guarantee a sustainable environmental management.

In this context, aquatic biological communities (benthic macroinvertebrates, water plants, fish) are the objects of safeguard actions and at the same time they are the markers of the health of water bodies.

People are becoming more and more aware that only the safeguard of the ecological integrity of rivers could guarantee the use of water in view of the sustainability and renew ability of this resource.

Rivers are generally heavily exposed to loads of polluting substances that can come from punctual (sewers, effluents of purification plants) or diffuse discharges (drainage of agricultural lands). In order to evaluate the quality of running waters it is necessary to take into account many chemical and physical parameters, such as pH, temperature, electrical conductivity, calcium and magnesium concentrations, total hardness, alkalinity, suspended matter, dissolved oxygen, organic and nitrogenous substances, COD (chemical oxygen demand), BOD_5 (biological oxygen demand), etc.

Chemical and bacteriological analyses, however very important, cannot yield enough information on the whole of the river ecosystem. There are many variables in this kind of ecosystem, from the head to the mouth of the river: flow, turbulence, typology of sediments, chemical composition, structure of the river beds (whether they are straight or winding, wide or canalized), the riparian ecosystem populating the riverbanks and functioning as a filter or buffering zone, in that it retains and transforms, both physically and biologically, sensible levels of substances drained from the surrounding territory. In some instances, chemical analyses may not detect the presence of a given contaminant due to the dilution phenomenon. Hence, the importance of using bioindicators able to provide information integrated over time.

Table 1: Water-related selected indicators of EEA and UNEP reports [2, 3].

EEA	UNEP
Physical and Chemical Water Quality	
Water transparency in lakes; Nutrients in waters; Discharges of organic matter from point sources; Use of fertilizers; Number of livestock; Consumption of pesticides and pesticides in surface water and groundwater; Heavy metals in rivers; Hazardous substances in lakes and rivers; Accidental, legal and illegal discharges of oil; Legislative compliance.	Biochemical oxygen demand in water bodies; Releases of nitrogen (N) and phosphorus (P) to coastal waters; Discharges of oil into coastal waters.
Biological and Ecological Water Quality	
Development of urban waste water treatment; Legislative compliance; Chlorophyll in coastal and marine waters; Loss of habitats; Non-indigenous species; Environmental impact of fishing; Biological effects of hazardous substances on aquatic organisms; Hazardous substances in marine organisms; Legislative compliance.	Wastewater treatment coverage; Concentration of faecal coliform in freshwater; Algae index; Maximum sustained yield for fisheries.
Water Quantity Indicators	
Groundwater levels; Available water; Water exploitation index; Total water abstraction; Water consumption index; Sectorial use of water; Agricultural water use; Water use by households; Overall reservoir stocks; Saltwater intrusion; Water prices; Water use efficiency; Water leakage.	Groundwater reserves; Annual withdrawal of water; Domestic consumption of water per capita; Density of hydrological networks; Population growth in coastal areas.

The biomonitoring natural fresh running water environments affects the use of indicator species or communities. Generally benthic macroinvertebrates, fish, algae and a number of water plants [4] are used. However, macroinvertebrates are the most frequently used [5–8]. By macroinvertebrates we mean all invertebrate organisms – adults or larvae – bigger than 1 mm, living in streams and rivers. Among the most important: Insects, Crustaceans, Molluscs, Hirudineans, Triclads, Oligochaeta, etc.

The presence or absence of an indicator species (or community) reflects the environmental conditions and can be considered as one of the first signals regarding the health of the water body under study. However, a given species could be absent for reasons other than the presence of pollutants, such as for instance predation, competition or the possible presence of geographical barriers preventing them from entering a site. Biomonitoring methods for water classification according to the directive 2000/60/CE are ranked as follows:

1. methods based on the composition and abundance of macrobenthic communities;
2. methods based on composition, abundance and age structure of water fauna;
3. methods based on composition and abundance of water flora (macrophytes, phytoplankton and periphyton);
4. methods based on the whole river ecosystem.

2.1 Methods based on macrobenthic communities

Communities of benthic macrovertebrates undergo qualitative and quantitative alterations in response to possible chemical and physical events and to hydraulic and morphological changes of the water body concerned (fig. 2). Several indices are used to test the effects of contaminants on aquatic populations and communities.

Biotic indices are generally typical of the kind of contamination and of the geographical area; they are utilized to classify the level of contamination by determining the tolerance of an indicator organism to the polluting agent. The indicator species are classified according to their tolerance levels. These indices assume

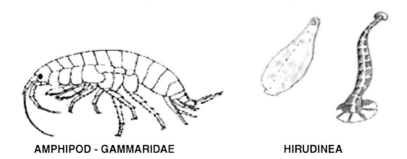

AMPHIPOD - GAMMARIDAE   **HIRUDINEA**

Figure 2: Examples of macroinvertebrates.

that the system or the polluted sites have a lower number of species in comparison with the uncontaminated sites or ecosystems. The measurements generally concern indices such as richness, tolerance to contamination, trophic levels and abundance. In particular, these indices are used for the evaluation of organic contamination.

Diversity indices estimate the number of different taxa (for instance: orders, families, species) present in a site, as well as uniformity, the relevant abundance of different taxonomic groups; their number is determined by counting sampled organisms. The absence of groups of benthic macroinvertebrates sensitive to contamination (ephemeroptera, plecoptera and trichoptera) and the dominance of groups tolerant to contamination (oligochaeta or chironomidae) is indicative of a scarce quality of the water body.

Comparison and similarity indices evaluate the structure of a community in terms of richness and/or uniformity over time and space. An example of a comparison index is the index of biotic integrity (IBI). This index compares the degree of distribution and abundance of different species of fish in the water body in comparison with reference values of the same ecosystem (see below). There is a similar index, the index of the community of invertebrates (ICI), comparing distribution and abundance for macroinvertebrates.

The Trent Biotic Index was proposed in 1964 and later updated as Extended Biotic Index (EBI) [9] and subsequently adapted for a standardized application to Italian streams [10, 11].

This method is based on the study of the composition of communities of macroinvertebrate organisms present in the different stretches of the water body. At present, the EBI [11] is one of the most common indices and it is based on the richness in taxa of the communities and on the different sensitivity of some animals to contaminants. It is quite widespread in Italy and it is considered by the Ministerial Decree regulating the water legislation (DL 152/99) as a fundamental method of analysis for monitoring superficial waters; it is one of the parameters in the grid for the definition of the ecological status of the water body.

This method is suitable for the monitoring of superficial running waters and it has a good synthesis capacity. However, it is characterized by a low analysis capacity and therefore it must be backed by the chemico-physical and bacteriological data of the waters in order to identify the causes of pollution.

One of the biggest problems in the evaluation of the ecological integrity of water bodies is the lack of information on the macrobenthic communities of references with which the communities under scrutiny are compared [12].

Other methods take into consideration mainly the functional aspects, and rather than measuring taxa (e.g. species, genus, family). They measure qualitative and quantitative abundances of the various trophic-functional categories (herbivores, detritivores, carnivores) that are always compared to reference values and utilized as a measurement of the integrity of the community. Table 2 reports the advantages and disadvantages of the use of benthic macroinvertebrates; table 3 shows the presence of macroinvertebrate communities as a response to a given impact on the water body.

Table 2: Advantages and disadvantages in the use of benthic macroinvertebrates (modified from Rosenberg and Resh; Plafkin *et al.*; Klemm
et al. [5, 13, 14]).

Advantages	Disadvantages
They are available in many water habitats.	They are not affected by all environmental impacts.
They include numerous species.	They are sensitive to seasonal cycles.
Numerous communities are present also in small streams.	They are easily transported by the water currents.
They have limited mobility.	They prefer rocky habitats.
Easy to identify; their sampling requires minimum equipment.	
Changes in their communities indicate a certain degree of environmental deterioration.	
The negative factors affecting them have repercussions on the food chain.	
The toxic substances accumulated are easy to analyse.	
Long life cycles (one year or more), steadily present in the water body.	

Table 3: Response of macroinvertebrate communities to pollution (modified from
De Pauw and Hawkes [6]).

Kind of deterioration	Increase (+) or decrease (−) of invertebrates
low concentration of dissolved oxygen	+ aquatic worms (oligochaetae) + midges (chironomidae)
nutrient increase	+ water worms (oligochaetae) + midges (chironomidae) + ephemeroptera
presence of heavy metals	+ aquatic worms (oligochaetae) + midges (chironomidae) + predators + bugs and water beetles
sedimentation	− chironomidae and ephemeroptera
low pH	− snails − bivalve molluscs − mussels − daphnias − ephemeroptera − chironomidae

Table 4: Obligatory limits for the definition of Systematic Units (SU) (from Ghetti [11]).

Animal groups	Taxonomic determination levels to define SU
Plecoptera	Genus
Trichoptera	Family
Ephemeroptera	Genus
Coleoptera	Family
Odonates	Genus
Diptera	Family
Heteroptera	Family
Crustaceans	Family
Gastropoda	Family
Bivalves	Family
Tricladida	Genus
Hirudinea	Genus
Oligochaeta	Family
Megaloptera, Planipennis, Nematomorphs, Nemertina[a]	Presence

[a]Taxa not frequently found in Italian running waters.

In order to determine the EBI, a special close-meshed net (less than 1 mm) is used to collect macroinvertebrates in the stretch of river under study. Then animals are classified and separated, thus allowing the identification of the Systematic Units (SU) present (the different Genera or Families). This index requires a systematic identification of macroinvertebrates down to the taxonomic level – established experimentally – of genus or family which varies with to the group and is useful to define SU (table 4) [11].

If the river is highly polluted there is a decrease in animal biodiversity (or a smaller number of SU). As already said, the main problem is finding valid information on the communities of reference, which, by definition, should refer to that very same site at a time when there was no presence of phenomena due to the anthropic impact.

The qualitative responses of the communities are closely linked to the different sensitivity of organisms to pollution. The species disappear, starting with the most sensitive (plecoptera, trichoptera and ephemeroptera) up to the strongest (Diptera larvae) (fig. 3).

The sampling technique involves collecting various taxa with a net provided with handle and following oblique transects diagonally to the river. After an initial *in vivo* analysis, the organisms are transported to the laboratory for a definitive classification with the help of stereo and transmission microscopes and classification keys.

The calculation of the value of the IBE index takes into account sampled organisms based on their sensitivity to pollution and the amount of taxa of the community.

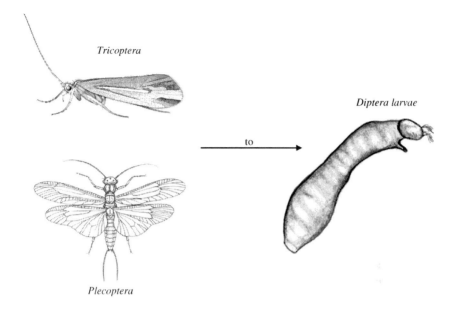

Tricoptera

Diptera larvae

to

Plecoptera

Figure 3:   Different sensitivity of organisms to pollution, from the most sensitive
(Plecoptera, Trichoptera) up to the strongest (Diptera larvae).

The IBE is calculated with the help of a two-input table (table 5) expressing
vertically the total number of SU collected and horizontally the decreasing degree
of sensitivity to pollution of the various taxa. The horizontal and vertical inputs
are chosen to correspond, respectively, with the most sensitive taxon sampled in a
given site and the total number on SU present; in their meeting point they indicate
the index value [11].

There are values ranging from 0 (highly polluted environment) to 12 (high quality environment). IBE values are converted into 5 quality classes, for each of which
there is a synthetic judgment and a conventional colour for the elaboration of the-
matic charts representing a given hydrographic grid (table 6). With these it is easy
to read charts and possible to immediately assess the environmental quality.

However, it is important to clarify the meaning that should be attributed to the
synthetic judgments defining each quality class (each represented by a colour):
these evaluations are correlated with the quality of the biological environment and
are not linked, at least not directly, to the different human use (drinking, bathing,
irrigation, industrial use). On the other hand, it has a direct correlation with the
safeguard of aquatic life and hence the environmental use of rivers.

Biomonitoring activities, backed by physico-chemical examinations allow
the planning of anthropic interventions on the river, evaluating its self-purifying
capacities; it can be used on the whole length of the river or on some specific
stretches, in order to spot illegal discharges or to study the effects of the impact
of authorized releases. This method, contrary to chemical analyses, makes it

Table 5: Table for the calculation of IBE value (from Ghetti [11]).

Animals	SU collected	Total SU forming the community								
		0–1	2–5	6–10	11–15	16–20	21–25	26–30	31–35	36
Plecoptera (Leuctra)[b]	More than one	—	—	8	9	10	11	12	13[a]	14[a]
		—	—	7	8	9	10	11	12	13[a]
Ephemeroptera (Baetidae Caenidae)[c]	More than one	—	—	7	8	9	10	11	12	—
		—	—	6	7	8	9	10	11	—
Trichoptera	More than one	—	5	6	7	8	9	10	11	—
		—	4	5	6	7	8	9	10	—
Gammaridae, Atyidae and Palaemonidae	Above absent	—	4	5	6	7	8	9	10	—
Asellidae	Above absent	—	3	4	5	6	7	8	9	—
Oligochaeta or chironomidae	Above absent	1	2	3	4	5	—	—	—	—
Taxa absent above	Above absent	0	1	—	—	—	—	—	—	—

– Uncertain judgment due to sampling error, presence of drift organisms not excluded in the calculation, environment not suitably colonized, typology not valuable through the index (e.g. sources, snowfield melt waters, stagnant waters, deltaic areas, brackish areas).

[a] These index values are rarely found in Italian running waters, therefore it is important to be very careful to avoid the sum of biotypology (artificial increase in the number of taxa) and when evaluating possible side-effects caused by pollution, since these environments are naturally rich in taxa.

[b] In communities where Leuctra is the only Plecoptera taxon and Ephemeroptera are absent (or where only Baetidae and Caenidae are present), Leuctra must be considered at the same level of Trichoptera in order to determine the horizontal input in the table.

[c] For the definition of the horizontal input in the table, Baetidae and Caenidae are considered at the same level of Trichoptera.

Table 6: Conversion of IBE values into Quality Classes (QC), and relevant judgment and standard colour for cartographic representation (from Ghetti [11]).

QC	IBE	Judgment	Colour
I	10–11–12	Non-polluted or non-sensibly altered environment	Blue
II	8–9	Environment showing some pollution effects	Green
III	6–7	Polluted or altered environment	Yellow
IV	4–5	Highly polluted or highly altered environment	Orange
V	1–2–3	Very highly polluted or very highly altered environment	Red

possible to register variations in the microfauna caused by the impact of man-made works (embankments, dykes). These works, although they are not themselves sources of pollution, create alterations in some of the river properties (light, flow variations) and therefore they cause limitations to the lives of macroinvertebrates.

However, it is necessary to have reliable quantitative information on reference levels, such as historical data, natural background values, data of a geographical reference ecosystem [15].

As reported above, macroinvertebrate taxa distribution and density are influenced by many factors such as organic pollution, pesticides and habitat degradation [16]. In this view, the use of multivariate techniques have been proven to be a potentially useful tool when dealing with ecotoxicological problems [17]. In fact it allows testing data taking into account the global contribution of all the variables in the data set [18].

In particular, the canonical correspondence analysis (CCA) showed good results for the study of the effects of environmental conditions on species assemblages. CCA seems to be superior to redundancy analysis (RDA) in that CCA assumes that ecosystems variations occur along non-linear gradients. This is probably the best way to have a description of a multivariate ensemble of species in environmental space [16].

Berenzen et al. [16], by using both CCA and RDA, showed that the toxicity of pesticides mixtures (studied by the toxic unit – TU approach) in some German streams made the greatest contribution toward explaining the variance in macroinvertebrate communities.

2.2   Methods based on fish fauna

Fish can be used as indicators of ecological integrity of running waters [19], especially for non-migratory species, such as pike and perch, which can be used, for instance, for an accurate evaluation of the presence of mercury and chlorine organic compounds in water [4]. Fish can also be used as biomarkers of the quality of the ecosystem (see Chapter 2).

Fish mortality is partly due to environmental factors (temperature, dissolved oxygen and water variations) and partly to predation and food scarcity; as a matter of fact, each animal species only populates a given environment if food, reproduction and shelter conditions are suitable – at least to a certain degree – to conduct their life cycle. All running water fish species are suited – from an evolutionary point of view – to live in river stretches presenting physico-chemical characteristics and water trophism.

Therefore, they reflect environmental conditions both through the loss of species and the modification of the population which can be affected by scarce reproductive success or slow growth.

They are relatively easy to collect and identify and, besides, the evolution and distribution of some of the species is well known. They are good bioindicators of long-term (several years) environmental impact and occupy various trophic levels.

On the other hand, the qualitative evaluation of a water body cannot consist solely in the identification of the animal species, since fish do not have the same sensitivity to pollution as macroinvertebrate communities. Moreover, some animal communities are strongly altered by the introduction of exotic species that could be able to alter considerably the structure of the community, as happened with Crucian carp and Wels catfish in Italian rivers [12].

As stated by Fausch et al. [20], the general approach for the study of environmental degradation regards the following aspects:

1. the use of indicator taxa or guilds;
2. indices of species richness, diversity and evenness;
3. multivariate methods;
4. the IBI.

Generally, researchers select indicator taxa or guilds on the basis of their declining abundance with the environmental degradation. This method is simple and can be applied with relative abundance or presence/absence sampling of fish communities. This approach can be considered indicative because it has many uncertainty problems. For instance, it is difficult to confirm whether a species is sensitive to a specific pollutant solely on the basis of empirical evidence from the field [20]. Moreover, taxa could be absent for other reasons than pollution phenomena, such as biogeographic barriers, biological interactions, over harvesting. Besides, a single species used to measure alterations at the level of the individual organism and population cannot be used as a robust measure at ecosystem level [20].

The richness of a species (number of individuals in a sample from a community) is the most important ecological attribute of a community [21]. The relative abundance (evenness) and richness and also the diversity have been widely used to assess the structure of communities [22].

For measuring diversity Shannon–Wiener and Brillouin are the most commonly used indices. The Shannon–Wiener index H' is a diversity index used in statistics for populations with an infinite number of elements:

$$H' = \sum_{j=1}^{s} p_j \log p_j$$

where s is number of species in a sample from the community; p_j the relative frequency of each species; it is expressed as the ratio n_j/N (n_j number of individuals in a sample from the community, N total number of individuals in the community).

One of the most frequently employed indices in the environmental field, it expresses the degree of relative uniformity in the numerical strength of all sampled species. It measures the mean uncertainty value when postulating the species of an individual randomly picked from a collection of data [23]. It requires that individuals be randomly picked and that all species be actually represented in the sample. It assumes values ranging between zero – in the presence of one single species – and the logarithm S of the totality of the species present, when all species are represented by the same number of individuals.

The Brillouin index H_B is a diversity index used in statistics for populations with a finite number of elements.

$$H_B = \frac{\ln(N!) - \sum_{i=1}^{s} \ln(n_i!)}{N}$$

This index can be employed when the entire population has been collected or when the randomness of the sample cannot be guaranteed (for instance, when the species are attracted or caught by the sampling apparatus in a non-homogeneous way) [23]. This index has less advantages in comparison with the Shannon–Wiener index, since the Brillouin index cannot give information on uncertainty: it describes an unknown collection and is strongly influenced by sample size [20, 23].

The evenness index is

$$E = \frac{H}{\ln S}$$

where H is the Shannon–Wiener index and S is the number of species. This index expresses, in an interval ranging from 0 to 1, the degree of evenness of relative abundances (dominance) of the species, allowing the distinction between the input of the species variety from the input of their dominance of the calculation of general diversity.

For example, Shannon–Wiener indices varied significantly between seasons, indicating meaningful changes in fish species composition for subtropical streams studied in Brazil [24].

Also multivariate analysis has been widely employed to evaluate the environmental degradation using fish communities as bioindicators. In general, multivariate methods have the advantage of the quantitative data elaboration that allows the comparison for both community structure and functions; it allows simultaneous comparison between all samples. On the other hand, these procedures are complicated to read; besides, results highly depend on the reference samples used [18, 20].

For European streams, the presence of endemic species allows the identification of stretches reasonably well defined from an ecological point of view, and therefore the species under study can be considered to be good bioindicators.

For example, the presence of some species, especially during the breeding season, in a river stretch or in a lake area, means that the water has a set of physico-chemical characteristics, temperature, quantity of dissolved oxygen, saline composition and conductivity and therefore it is indicative of the absence or moderate presence of toxic substances [25].

For European, and particularly Italian rivers, Zerunian [26] suggested that river stretches be zoned on the basis of the fish species present.

2.2.1   Trout zone
It is the upper stretch of a river where water runs fast and is rich in oxygen, the bottom is made of big gravel and therefore the water is very clear and temperature is always below 13°C.

2.2.2   Zone of cyprinidae – lithophilic deposition
This fish class is constituted mainly by Barbel and Chub. In this area, water generally runs fast as opposed to other stretches where the flow could be slower due to depth.

Here the river bed is made of fine gravel, therefore the water turns out to be less clear than the previous stretch.

Temperatures are about 12–20°C.

2.2.3   Zone of cyprinidae – phytophilic deposition
This area is comprised of the middle and lower stretches of a river, where water is deep and rather turbid, the bed is mainly muddy and the flow is slow.

Temperatures can be as high as 22°C and there is the presence of macrophytes.

The fish that can be found here are Tench and Rudd.

2.2.4   Brackish zone
Brackish zone is the stretch close to the mouth of the river, a zone where fish like the Flounder and the Grey mullet live. These fish are able to adapt to fresh and salty water alike, thanks to their osmotic conditions.

However, the scarce knowledge of the ecology of many fish species in European rivers is a serious hinder towards a correct classification of all the endemic species of each river stretch. Another problem is the constant introduction of exotic species as well as the high rate of interspecific hybridization which make the reading of the collected data difficult.

According to the Environmental Protection Agency the fish community can reflect the overall health status of the surrounding environment, since fish are present in numerous habitats, have a relatively long life and occupy different trophic levels within a community (omnivores, herbivores, insectivores, planktivores, piscivores). Moreover, they allow an evaluation of the condition of the river integrated over a wide time interval and show effects due both to acute toxicity (loss of systematic units) and sub acute toxicity (for instance, reduced growth rate and declining reproductive success). Besides, fish are relatively easy to sample and their specific determination can be carried out in the field without causing harm to animals.

Among the most commonly used methods for the evaluation of the quality of the rivers we have the Rapid Bioassessment Protocol V-Fish [13, 27, 28] based on Index of Biological Integrity (IBI) [21].

This method involves the sampling of fish fauna through the use of an electric stunner in a representative stretch of the river that is being studied. The fish collected are counted, determined at specific level, the percentage of omnivorous species, of insectivorous and piscivorous cyprinidae, of hybrid individuals and individuals in bad conditions (tumours, parasite disease, skeletal anomalies, etc.) is evaluated. Scores (1, 3 or 5) are assigned for each variable (12 in all). The sum of these scores ranges from 12 to 60.

2.3   The Index of Biotic Integrity

The Index of Biotic Integrity (IBI) [21] is characterized to evaluate the ability of a biological community to remain balanced and suitably integrated in the ecosystem under study.

IBI requires the calculation of three groups of variables relating to (1) composition and specific richness, (2) trophic structure, (3) abundance and health of fish fauna.

The variables of the first group, six in all, include the definition of the total number of species, the number of species and identity of particular taxonomic groups and the number of intolerant species.

The variables concerning the trophic structure (second group) are: percentage of the presence of omnivorous fish, insectivorous cyprinidae and piscivores. The third group includes overall abundance, the percentage of hybrid individuals and of individuals in bad conditions (diseases, tumours, skeletal anomalies, etc.).

For each one of these variables, also as a function of the dimension of the river and of the geographical region, modalities of variation are defined: for instance, the increase in the percentage of omnivorous fish is proportional to the decline of the environmental quality and therefore also of biotic integrity; then, threshold values are identified, defining three different integrity classes for each group of variables.

Therefore, each group of variables is given a score:

1. 5 if no deviation from the expected values is detected for a site in natural conditions;
2. 3 if deviations, however not very noticeable, are detected;
3. 1 if deviations are very noticeable.

The total sum will constitute the *site score*, which will in turn fall in one of the five classes of biotic integrity (excellent, good, fair, poor, very poor). Table 7 shows

Table 7: Classes of biological integrity, their attributes and corresponding total index of biotic integrity (IBI) scores based on the sum of 12 metric ratings (from Fausch et al. [20]).

Integrity class	Attributes	Total IBI score
Excellent	Comparable to the best habitats without human disturbance; all species expected for the region, stream size, and habitat, including the most intolerant forms, are present with a full array of age- (size-) classes; balanced trophic structure and few hybrids or diseased fish; very few or no introduced species	58–60
Good	Species richness somewhat below expectation, especially due to loss of the most intolerant forms; some species have less than optimal abundance or age and size distributions; trophic structure shows some signs of stress, but still few hybrids or diseased fish; the proportion of individuals that are members of introduced species is usually low	48–52
Fair	Signs of additional deterioration include loss of intolerant forms, fewer species, and generally reduced abundances; skewed trophic structure is indicated by increasing frequency of omnivores and tolerant species (e.g. green sunfish *Lepomis cyanellus)*; older age-classes, especially of top carnivores, may be rare; incidence of hybrids and diseased fish increases above natural levels; the proportion of individuals that are members of introduced species increases	40–44
Poor	Community dominated by few species, many of which are omnivores, tolerant forms, habitat generalists or introduced species; few top carnivores; many age-classes are missing, and abundance, growth and condition are commonly depressed; incidence of hybrids and diseased fish is moderate	28–34
Very poor	Few fish present, most of which are introduced or tolerant omnivorous species; hybrids are common, and incidence of disease, parasites, fin damage and other anomalies is high	12–22
No fish	No fish are captured, even after repeated sampling	

classes of biological integrity and their attributes (proposed by Fausch *et al.* [20]). The lowest class, presenting a very poor biotic integrity, is characterized by the presence of few fish (mainly introduced and tolerant) of the hybrid kind and individuals in bad conditions.

On the other hand, the conditions of the fish community presenting an excellent biotic integrity (the class with the highest environmental quality) are comparable with the best condition that can be met with when human disturbance is absent.

The use of fish communities as bioindicators is a valid tool for the evaluation of the quality of rivers. However, it must be taken into account that the use of questionable ichthyogenic practices has sometimes seriously compromised the composition of communities endemic in freshwaters. In IBI, the evaluation of the biotic integrity also includes biogeographical variability.

As already said, the introduction of exotic species can modify meaningfully the structure of original communities.

In Europe, several biotic indices have been suggested which are a result of IBI. In Rumania, the preliminary multimetric indices (PMIs) [29]; in France a modified IBI [30] based on 10 fish assemblage attributes correlated to richness, specie and trophic composition, health and biomass. The scores obtained are based on regional standards. Also in Belgium the modified IBI has been suggested, including biomass among its parameters [31]. IBI was also adapted for Lithuanian rivers; it too includes biomass among its parameters [32]. In Italy, some authors [33] have proposed an index to be used on carp-like fish in river areas. Fish material must be collected, using an electrical stunner on an area of approximately 100 m².

It is a numerical index based on the composition and structure of the fish community. It is comprised of 5 quality classes in comparison with a reference community and it does not differentiate between allochthonous and autochthonous species. It has been tested in Apennine rivers in the Piedmont region (Northern Italy) [33].

The Ichthyological Index (I.I.) summarizes the response of the community in its trophic structures to changes due to pollution.

The index is obtained (table 8):

$$\text{I.I.} = S + a + b;$$

where S is number of sampled species on a standard area of 100 m², a the correction factor based on the number and classes of size of the species for sample and b the correction factor based on the number of sampled d species.

The importance using multimetric indices (i.e. IBI) is due to two factors:

1. they have a large ecological meaning because they respond to different kinds of impact (hydrologic regime, habitat configuration, biotic integrations);
2. they are part of a "regional framework" (Ichthyogeographic regions).

Table 8: Classification of rivers on the basis of Ichthyological Index (from Lodi and Badino [33]).

I.I.	Class	Colour	Fish community
>18	I	blue	the majority of the characteristic species present in balanced percentages
8–17	II	green	many species are still present, but there are some typical species which are absent
5–7	III	yellow	seriously damaged fish community
2–4	IV	orange	unbalanced structure of the community, only a few species present
<1	V	red	small or totally absent community

Multimetric indices must be duly validated experimentally before they can be used. Candidate metrics are assessed for responsiveness to biotic or abiotic factors resulting from increasing anthropic stress and their biological relevance [34].

The parameters of the index must be validated, resorting, for instance, to the Site Quality Index (SQI), based on hydrologic and physic alterations, chemical quality of waters, biological characteristics of the species (origin, trophic characteristics, reproductive strategies, sensitivity to pollution, etc.). SQI requires the evaluation of 32 possible parameters for which the degree of association and redundancy will be calculated [29].

2.4   Methods based on plant communities

Water vegetation can yield important information on environmental quality. Taxa can be used as an indicator of the trophic level of water and it is known that some oligotrophic species disappear when pollution increases or that the presence of some plants grows proportionally with the growth of nutrients in the water body. Methods employing plants as bioindicators use indices based mainly on water macrophytes and on the vegetation of the river ecosystem [35, 36].

Macrophytes include numerous macroscopic species. They are mostly herbaceous Phanerograms, some Pteridophytes, numerous Bryophytes (Hepatic and Mosses) and macroscopic algae. A specie is considered to be a bioindicator when the variation of its frequency of abundance is statistically correlated with ecological changes.

Factors to be taken into account when studying macrophytes are:

– abiotic factors: they can be of a morphometric or chemical nature.

Morphometric parameters are:

1. the mean height of the water body;
2. speed rate of flow, which is estimated according to the elaboration of a simple sequence (very fast, fast, intermediate, slow, very slow)
3. the sediment granulometry;

4. depth of the water body;
5. light, which is ranked in 5 classes (very bright, bright, intermediate, dark, very dark).

The most commonly studied chemical parameters are:

1. the determination of the mineral component of water and other parameters: hardness (determination of Ca and Mg), total alkalinity, sodium, potassium, phosphates, chlorides and conductivity;
2. the evaluation of the trophic level of water through the determination of orthophosphates, total phosphorus, nitrates, nitrites, COD, BOD_5, pH, dissolved oxygen. These last two parameters are very variable over time and must be accurately correlated with sampling time.

- biotic factors: biological cycle, competition, grazing;
- anthropic factors: morphological or hydraulic alterations, pollution, eutrophication, introduction of exotic species.

It is very important that the macrophytic indices described above be obtained through strict and homogeneous sampling methods. They have in common most of the survey procedures, which are defined by the Guidance Standard for the surveying of aquatic macrophytes in running water [37]. According to the standard procedure, macrophytes of a standard length present in a river stretch must be observed, identified and registered. Then, the total and specific covering must be accurately estimated. The environments to be sampled (immersed macrophytes, over aquatic zone) and the right time for the sampling (that is, during the growing season, when the water is low) are also indicated.

As far as identification is concerned, the various methods require the determination to the genus for algae and to the species for the other systematic categories.

When making an inventory, macrophytes are to be divided according to the type of river, defining the reference community. A comparison with other indicators of status and relation with pressures is also useful.

Different indices use macrophytes as a way to measure the quality of water bodies; fundamentally, they are based on the presence/absence and abundance of some indicator species duly chosen according to their sensitivity to pollutants. Some methods involve a census of all species, each of which is assigned its right sensitivity values.

Macrophytic indices were first elaborated in the 1970s (Ireland, UK, France, Germany). These indices, generically focusing on the integrated evaluation of water quality and the integrity of the river ecosystem, have a very high sensitivity to organic pollution.

They are currently very widely used in many European countries, such as France, UK, Ireland, Germany, Luxembourg, Belgium, Austria, Spain, Portugal [38–43].

Indices can be of two kinds:

1. those based on an evaluation of presence/absence and (only in some instances) abundance of a limited set of indicator taxa ranked in sensitivity/tolerance

classes. Examples of this class are the Macrophyte Index Scheme (MIS) [44]. This index is based on the presence/absence and dominance of some taxa with different degrees of sensitivity (4 groups). One limitation is that it requires cenosis with highly developed macrophytes and it takes into consideration only 29 indicator taxa. Another index of the same kind is the Nutrient Status Order.

2. those based on the use of a greater number of indicator taxa, each one associated to a tolerance/sensitivity coefficient and, sometimes, an indicator value, also known as reliability value. The indices belonging to this class are the following: The trophic index macrophyten (TIM) [45] is a trophic index based on the species richness with a different indicator value. Waters are ranked according to seven different trophic levels. It takes into account a higher number of taxa in comparison with MIS. The Mean Trophic Rank (MTR) [46–49] expresses the impact of the phosphate concentration on the macrophytic community with a score going from 10 (maximum impact) to 100 (absence of impact). It is based on indicator values of 129 species and on the percentage of covering.

The Indexes *Groupement d'Intéret Scientifique* (GIS) [41] are an inventory of vegetable species between May and October on stretches measuring at least 50 m in length, aquatic and over aquatic zones. They combine the covering coefficient with an indicative score (0–10). Different indices are calculated for the different zones and referred to presence/absence, abundance/dominance and also based on their tolerance to degradation in polluted areas.

The aim of index *Biologique Macrophytique en Rivière* (IBMR) [50] is to determine the trophic status of natural and artificial rivers in the continental region. It combines covering percentages with a specific oligotrophic score and with a coefficient of adaptation to stressed areas. The trophic classification of the river is done according to 5 quality classes. Also in this instance, the estimate of the covering percentage must be very accurate.

Another widely used method to evaluate the quality of a water ecosystem is the multivariate analysis, with a description of abiotic parameters (see above) and of aquatic vegetation present in rivers [51, 52].

Generally, for each site, a river stretch measuring 50 m is chosen for the sampling of plants. Aquatic macrophytes can be found both in the outer and inner regions of superficial lotic and lenitic fresh waters. Then a qualitative-quantitative inventory is elaborated based on the percentage of the macrophytes present in the site that can include also filamentous algae, bryophytes and vascular plants.

By using statistical methods, as for instance PCA, a possible correlation between flora groups and environmental parameters is established. PCA can be supplemented with the hierarchical ascendant classification (HAC), which can further confirm the bioindication capacities of various species. HAC uses data yielded by PCA and ranks them in various classes each characterized by its specific scientific parameters and by their water flora [51, 52].

Macrophytes are a relatively limited number of species present throughout Europe, and this is one of the reasons why they are so widely used as bioindicators.

Bryophytes have a high capacity of absorption and accumulation of substances sampled from water and the atmosphere, since they have no cuticle or stomatal openings; therefore, gas exchanges take place through the whole of the plant surface. Quite often, bryophytes make up the greatest part of macrophyte vegetation of aquatic fresh water habitats [53]. For biomonitoring purposes of aquatic habitats, only bryophytes submerged throughout the year can be used. The various species of mosses and hepatics have different tolerance degrees to pollutants. They are generally used to monitor heavy metals [54–56] but also nitrogen, sulphur, toxic organic compounds and radionuclides (see Zechmeister *et al.* [53] for extensive review).

The plant accumulates contaminant in function of their concentration in the atmosphere or in water, and of time of exposure and consequently, with an equal concentration in the environment, the highest degree of contamination is in the oldest thallus. Therefore, bryophytes can be used as accumulative indicators or for the overall evaluation of the water quality (nutrient status or pH changes) through the study of bryophyte species assemblages [53].

It is possible to use native plants, that is plants that are naturally present in that aquatic ecosystem, or they can be transplanted from other sites (that can be contaminated or uncontaminated, depending on the kind of study that is being carried out).

Generally, the moss to be transplanted is collected in an uncontaminated zone; then it is washed with distilled water and put in special cubic rigid plastic mesh containers. These samples are put in water and fixed with ropes on the riverbanks.

Particularly important is the aquatic moss *Fontinalis antipyretica* that was analysed as a biomonitor to assess the heavy metal contamination of an industrial effluent discharge or to evaluate the bioaccumulation capacity of native and transplanted mosses [57–59].

In particular, transplanted mosses turned out to accumulate significantly more Al, Cr, Cu, Pb, V and Zn than the native mosses studied in five streams of Poland [59]. In these kinds of studies is also important to have information on pH of the studied rivers, because it highly affects the heavy metals uptake.

Methods based on periphytic communities concern complex communities of microscopic organisms living close to submerged substrata. Periphyton responds to ecosystem alterations and it can provide both structural and functional information.

Functional alterations concern the internal dynamics of the energy of the system and are based on the measurement of some activities such as primary production, respiration, photosynthesis, microbial enzyme activity and nutrient uptake.

Among structural parameters we find total biomass, autotrophic biomass and qualitative and quantitative analysis of the colonizing species. The integrated use of these two approaches allows a complete description of the response of the ecosystem to polluting events [12].

Among the most representative organisms of the autotrophic component of periphyton we find diatoms (or siliceous algae) which are the basis for biotic indices for the assessment of the river water quality [39, 60, 61].

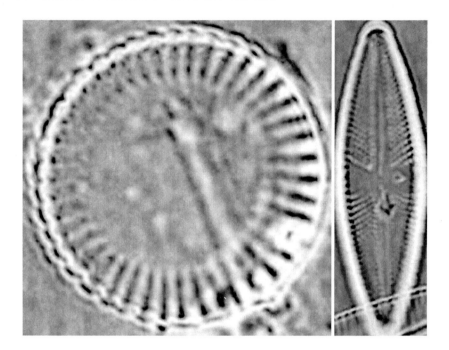

Figure 4: Diatoms Centrales (left) and Pennales (right).

There are two orders of diatoms: Centrales and Pennales (fig. 4). The first consists of radial symmetry cells, and the second of bilateral symmetry cells. The first order consists of planktonic forms, while in the second there are both planktonic forms and so-called benthic forms from the bottom.

The majority of diatoms are unicellular and live as single cells or combining in chains, thus enhancing their floating capacity. Their external skeleton, also called frustule, is composed mainly of silicon and is made of two valves separated by a longitudinal sheath on which there are tiny "teeth" allowing the two valves to adhere. The shape of the valves varies according to the species: it can be cylindrical, ellipsoidal, triangular, square and irregular. The valves also show various forms, sometimes quite complex of ornaments (i.e. spikes and teeth) that can be observed with the electron microscope and that are very important to distinguish the various species of diatoms.

There are forms of planktonic or benthic type living free in the water. The latter are practically the only ones populating running waters. They can be divided in three classes, according to their habitats:

1. epilithic, forming a thin brownish coating on submerged boulders, rocks and pebbles;
2. epiphytic, living on aquatic macrophytes or on macroscopic algae;
3. epipelic, settled on the mud at the bottom of the river bed, in the river stretches where waters flow slowly.

Many indices using diatoms are based on an equation taking into account abundance, that is the percentage of species in the sample and their sensitivity to pollution. Diatoms are considered to be good indicators because:

1. they include a high number of species with different ecological value;
2. they are sensitive to both organic pollution and salinity variations and particularly to chlorine and eutrophication;
3. they have rapid reproduction rates and short life cycles. Therefore, they are good indicators for short-term impacts.

Benthic diatoms are a useful tool for water quality assessment as well as macroinvertebrates and fish that integrate environmental quality on longer time-frames. Several diatom indices have been proposed in Europe to assess the biological quality of running waters. The more common indices proposed are:

Europe	CEE index – diatom Index of European Economic Community	[62]
France	Specific Polluosensitivity Index – SPI	[63]
	The Generic Diatom Index – GDI	[64]
	Standardized Biological Diatom Index – BDI	[65]
Belgium	Leclercq and Maquet Index – ILM	[66]
United Kingdom	Trophic Diatom Index – TDI	[39]
Germany	SHE index	[67]
Czech Republic	Sládeček index – SLA	[68]
Austria	Saprobic index of Rott – ROT `	[69]
Italy	Eutrophication Pollution Index using Diatoms – EPI-D	[70, 71]

Given their resistance to being dragged by the current, Pennales are considered more suited for the assessment of the quality of streams. Like for other groups of aquatic organisms, also diatoms have different species adapted to the most diverse ecological conditions. This favoured the conception of a method proposed by Dell'Uomo [70, 71] for the evaluation of the biological quality of Italian running waters. It is EPI-D (Eutrophication Pollution Index – Diatom based) and it is based on the sensitivity of the species to the concentration of nutrients and of organic substances and to the mineralization degree of the water body, with special reference to chlorides. In the monitoring of rivers mainly epilithic diatoms are used, since they are easily collected from the substratum with the use of a simple brush. For the application of this index, it is necessary to identify the species, and this is possible only after incinerating the organic matter and observing the preparations with a one thousand magnification optical microscope after applying a special resin with a high index of refraction.

According to Dell'Uomo [70, 71], the values of the EPI-D index range from 0 to 4; values close to 0 indicate clean waters, while the higher values indicate increasingly compromised waters; the results are ranked in 8 class qualities (table 9).

Table 9:   Eutrophication Pollution Index – EPI-D for Italian running waters (from Dell'Uomo [70, 71]).

EPI-D	Assessment	Colour
3.0–4.0	totally degraded water body	brown/black
2.5–3.0	very high pollution	red
2.2–2.5	high pollution	orange
2.0–2.2	moderate pollution	yellow
1.8–2.0	slight pollution	green
1.5–1.8	fairly good quality	dark green
1.0–1.5	good quality	blue
0.0–1.0	excellent quality	sky blue

The analysis of epilithic diatoms is to be carried out at least twice a year, during the seasons of low and moderate flow of the river.

2.5   Methods based on the whole of the river ecosystem

The ecological functionality of a stream concerns different aspects [72, 73]:

1. processes of macrobenthic colonization;
2. autochthonous and allochthonous food supply models;
3. cyclization and retention capacity of the organic compound;
4. trophic relations between living organisms (fish and terrestrial vertebrates, her-petofauna, mammal fauna and avifauna);
5. contribution to environmental diversity and biodiversity;
6. ecotonal functions;
7. the river as ecological corridor;
8. the river as effluent regulator;
9. the river as regulator of sediment transport, etc.

The whole river system constitutes a complex system comprising also river strips and portions of surrounding territory more closely interacting with the river.

Therefore, a stream can be considered as a succession of ecosystems gradually fading into one another: they are interconnected with surrounding terrestrial ecosystems. From source to mouth morphological, hydrodynamic, physical and chemical parameters change and, in connection with them, biological populations.

The River Continuum Concept [74] proposes a unifying vision of river ecology calling attention to how closely the structure and the function of a biological community depend on mean geomorphologic and hydraulic conditions of the physical system. Fig. 5 shows the proposed relationship between stream size and the progressive shift in structural and functional attributes of lotic communities [74].

The Fluvial Functionality Index (IFF) represents the gradual and continual evolution of new methods for the study and the control of streams, initially limited

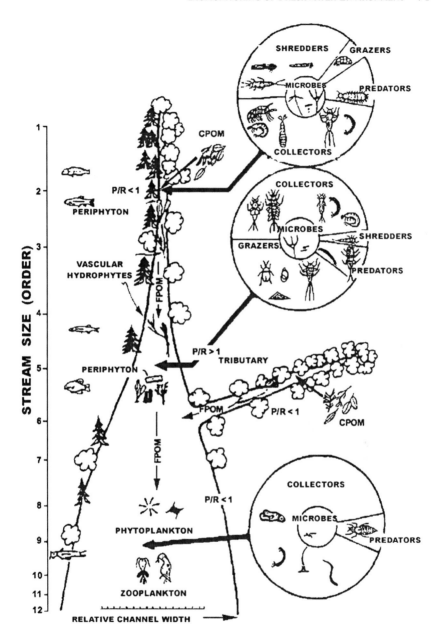

Figure 5: A proposed relationship between stream size and the progressive shift in structural and functional attributes of lotic communities (modified from Vannote *et al.* [74]). CPOM = Coarse Particulate Organic Matter; FPOM = Fine Particulate Organic Matter; P/R = Relationship Photosynthesis/Respiration.

to the evaluation of the sole chemical and microbiological characteristic of the water sample without taking into account the biological component of the whole ecosystem. At present, IFF takes into account the whole of the river ecosystem, differing in part from the indices based on single bioindicators such as microarthropods, macroinvertebrates and macrophytes. IFF originates from the RCE-I (Riparian Channel Environmental Inventory) proposed by Petersen (1992) with the aim of compiling an inventory of the characteristics of Swedish rivers. After an initial testing period, the method was modified and adapted to Italian conditions, under the name of RCE-II initially and then IFF [73], focusing mainly on the functional meaning of the characteristics found.

According to Siligardi *et al.* [73], the main components considered by the index concern:

1. vegetation in the area around the river: buffer strip (removal of nutrients and filter for sediments), consolidation of the banks and protection from floods, trophic input (leaves and branches), temperature regulation (shading), presence of habitat along the river corridor;
2. hydraulic regime: evaluation of efficiency and stability of the river bed colonization;
3. structure of the river bed and of the cross section: retention capacity of the organic matter, environmental diversity, creation of habitat for fish (in terms of macroscale) and benthic fauna;
4. vegetable component in the river bed: evaluation of the trophic status;
5. biological component: efficiency in the demolition of the detritus, evaluation of the structure and diversification of macrobenthic community.

After a preliminary study, following the stream from mouth to source, physically and morphologically homogeneous stretches are identified. For each stretch a record is compiled. The record is made up of 14 questions each having four possible answers associated to an accurately devised score [73, 75].

For each stretch and each bank an IFF value is calculated which is ranked in 5 levels of river functionality ranging from 1st (best conditions) and 5th (worst conditions). Each level has a corresponding functionality judgment and a conventional colour for cartographic representation (table 10).

Table 10: Classification of environmental quality (I–V) (from Siligardi *et al.* [73]).

Class	Score	Quality	Colour
I	251–300	Very good	Blue
II	201–250	Good	Green
III	101–200	Fair	Yellow
IV	51–100	Poor	Orange
V	14–50	Bad	Red

The sampling method requires highly specialized staff, with a good knowledge of river ecology, for an evaluation of the correct answer out of four possible answers to each question. Each answer corresponds to a particular condition of river functionality [73]. IFF is an excellent method to document the status of steams, stressing the impact of human intervention. Samplings are to be carried out in vegetation period and the surveillance monitoring is to be carried out for a period of one year for each monitoring site.

The described method is a rapid assessment to determine the management level of running waters in order to reach sustainable conditions due to the WFD requirements and it considers the watercourse in its entirety describing the relationship with its surrounding environment.

3   Conformity among groups of biological indicators

There are very few studies concerning conformity between the various biological indicators of lotic ecosystems. Ormerod *et al.* [76] have detected a strong conformity between macrophyte and macroinvertebrate populations in Welsh rivers according to an acidic gradient. The ecological impact of acidic emissions is a key topic on the control policy of lakes and streams [77].

Kilgour and Barton [78] studied the relationship between fish and benthic invertebrates communities in some streams in southern Ontario, Canada. They found good associations between fish and benthos.

However, Paavola *et al.* [79], in their research on 32 small Finnish rivers, have detected no conformity between fish, macroinvertebrates and bryophytes. The fish community is more affected by the stream depth, substrate size and dissolved oxygen (DO). Macroinvertebrates are more affected by "landscape-scale" factors, such as the size of the river [80] and of its basin and pH. Bryophytes turn out to be affected by factors such as water colour, nutrient content and in-stream habitat variability. In conclusion, great care should be taken when applying systematic group river classification to other groups.

References

[1] Directive 2000/60/EC of the European Parliament and of the Council of 23 October 2000 establishing a framework for Community action in the field of water policy. Official Journal L. 327 of 22.12.2000.

[2] Nixon S., Trent Z., Marcuello C., Lallana C. (2003) Europe's water: An indicator-based assessment. European Environmental Agency, Topic Report EEA Copenhagen 1/93, p. 97.

[3] Consultative Group on Sustainable Development, Dashboard of Sustainable Development Dataset (2002) Consultative Group on Sustainable Development Indicators, 9 January.

[4] Phillips D.J.H., Rainbow P.S. (1993) Biomonitoring of Trace Aquatic Contaminants. Elsevier Applied Science, New York.

[5] Rosenberg D.M., Resh V.H. (Eds.) (1993) Freshwater Biomonitoring and Benthic Macroinvertebrates. Chapman & Hall, Inc., New York, p. 488.

[6] De Pauw N., Hawkes H.A. (1993) Biological monitoring of river water quality, in: Walley W.J., Judd S. (Eds.), River Water Quality Monitoring and Control. Current Practices and Future Directions. Aston University Press, Birmingham, U.K., pp. 87–111.

[7] Cao Y., Bark A.W., Williams W.P. (1996) Measuring the responses of macroinvertebrate communities to water pollution: a comparison of multivariate approaches, biotic and diversity indices. Hydrobiologia, 341, 1–19.

[8] Clarke R.T., Wright J.F., Furse M.T. (2003) RIVPACS models for predicting the expected macroinvertebrate fauna and assessing the ecological quality of rivers. Ecological Modelling, 160(3), 219–233.

[9] Woodiwiss F.S. (1978) Biological Water Assessment Methods. Severn Trent River Authorities, United Kingdom.

[10] Ghetti P.F. (1995) Indice Biotico Esteso (IBE). Notiziario dei Metodi Analitici, pp. 1–24.

[11] Ghetti P.F. (1997) Indice Biotico Esteso (IBE) I macroinvertebrati nel controllo della qualità degli ambienti di acque correnti. Trento: Provincia Autonoma di Trento, Agenzia Provinciale per la protezione dell'Ambiente.

[12] Conti M.E. (2002) Il monitoraggio biologico della qualità ambientale, SEAM, Roma, p. 180.

[13] Plafkin J.L., Barbour M.T., Porter K.D., Gross S.K., Hughes R.M. (1989) Rapid bioassessment protocols for use in streams and rivers: Benthic macroinvertebrates and fish. EPA 440/4–89/001. Environmental Protection Agency, Washington US.

[14] Klemm D.J., Lewis P.A., Fulk F., Lazorchak J.M. (1991) Macroinvertebrate field and laboratory methods for evaluating the biological integrity of surface waters. Govt Reports Announcements & Index (GRA & I), Issue 14.

[15] Lorenz C.M. (2003) Bioindicators for ecosystem management, with special reference to freshwater systems (Chapter 4), in: Markert B.A., Breure A.M., Zechmeister H.G. (Eds.), Bioindicators & Biomonitors, Principles, Concepts and Applications. Elsevier, Amsterdam, pp. 123–152.

[16] Berenzen N., Kumke T., Schulz H.K., Schulz R. (2005) Macroinvertebrate community structure in agricultural streams: impact of runoff-related pesticide contamination impact of runoff-related pesticide contamination. Ecotoxicology and Environmental Safety, 60, 37–46.

[17] Sparks T.H., Scott W.A., Clarke R.T. (1999) Traditional multivariate techniques: potential for use in ecotoxicology. Environmental Toxicology Chemistry, 18, 128–137.

[18] Conti M.E., Iacobucci M., Cucina D., Mecozzi M. (2007) Multivariate statistical methods applied to biomonitoring studies. International Journal of Environment and Pollution, 29(1–3), 333–343.

[19] Chovanec A., Hofer R., Schiemer F. (2003) Fish as bioindicators (Chapter 18), in: Markert B.A., Breure A.M., Zechmeister H.G. (Eds.), Bioindicators & Biomonitors, Principles, Concepts and Applications, Elsevier, Amsterdam, pp. 639–676.

[20]   Fausch K.D., Lyons J., Karr J.R., Angermeier P.L. (1990) Fish communities as indicators of environmental degradation, in: Adams S.M., (Ed.), Biological Indicators of Stress in Fish. American Fisheries Symposium 8, Bethesda, Maryland, pp. 123–144.

[21]   Karr J.R. (1981) Assessment of biotic integrity using fish communities. Fisheries, 6(6), 21–27.

[22]   Washington H.G. (1984) Diversity, biotic and similarity indices – a review with special relevance to aquatic ecosystems. Water Research, 18(6), 653–694.

[23]   Moreno C.E. (2006) La vita e suoi numeri. Metodi di misura della biodiversità (a cura di M. Zunino). Bonanno Editore, Catania, p. 83.

[24]   Bozzetti M., Schulz U.H. (2004) An index of biotic integrity based on fish assemblages for subtropical streams in southern Brazil. Hydrobiologia, 529(1), 133–144.

[25]   Agenzia per la Protezione dell'Ambiente e per i Servizi Tecnici Centro Tematico Nazionale "Acque Interne e Marino Costiere" (2004) Indicatori biologici per i corsi d'acqua e canali artificiali, Firenze, p. 86.

[26]   Zerunian S. (1982) Una proposta di classificazione della zonazione longitudinale dei corsi d'acqua dell'Italia centro-meridionale. Bollettino Zoologia, 49(Suppl.), 200–216.

[27]   Karr J.R., Fausch K.D., Angermeier P.L., Yant P.R., Schlosser I.J. (1986) Assessing biological integrity in running waters: a method and its rationale. Illinois Natural History Survey Special Publication, 5, p. 28.

[28]   UNICHIM (1999) Linee guida per la classificazione biologica delle acque correnti superficiali. Manuale, 191, p. 59.

[29]   Angermeier P.L., Davideanu G. (2004) Using fish communities to assess streams in Romania: initial development of an index of biotic integrity. Hydrobiologia, 511, 65–78.

[30]   Oberdorff T., Porcher J.P. (1994) An index of biotic integrity to assess biological impacts of salmonid farm effluents on receiving waters. Aquaculture, 119(2–3), 219–235.

[31]   Belpaire C., Smolders R., Auweele I.V., Dirk Ercken D.E., Jan Breine J., Gerlinde Van Thuynel G., Ollevier F. (2000) An index of biotic integrity characterizing fish populations and the ecological quality of Flandrian water bodies. Hydrobiologia, 434, 17–33.

[32]   Kesminas V., Virbickas T. (2000) Application of an adapted index of biotic integrity to rivers of Lithuania. Hydrobiologia, 422/423, 257–270.

[33]   Lodi E., Badino G. (1991) Classificazione delle acque fluviali (zona di Cipriniformi) mediante l'Indice Ittico. Atti Accademia Scienze, Torino, 125(5–6), 192–203.

[34]   McCormick F.H., Hughes R.M., Kaufmann P.R., Peck D.V., Stoddard J.L., Herlihy A.T. (2001) Development of an index of biotic integrity for the Mid-Atlantic Highlands region. Transactions of the American Fisheries Society, 130, 857–887.

[35]   Carbiener R., Trémolières M., Mercier J.L., Ortscheit A. (1990) Aquatic macrophite communities as bioindicators of eutrophication in calcareous oligosaprobe stream waters (Upper Rhine Plain, Alsace). Vegetatio, 86, 71–88.

[36] Mériaux J.L. (1982) L'utilisation des macrophytes et des phytocenoses aquatiques comme indicateurs de la qualité des eaux. Naturalistes Belges, 63, 12–28.

[37] CEN. Guidance Standard for the surveying of aquatic macrophyte in running water – prEN 14184.

[38] Haury J., Peltre M.C. (1993) Interest and limits of macrophyte indices to qualify the mesology and the physical chemistry of rivers with examples in Armorica, Picardy and Lorraine. Annales de Limnologie, 29(3–4), 239–253.

[39] Kelly M.G., Whitton B.A. (1995) The trophic diatom index: a new index for monitoring eutrophication in rivers. Journal of Applied Phycology, 7(4), 433–444.

[40] Tremp H., Kohler A. (1995) The usefulness of macrophyte monitoring-systems, exemplified on eutrophication and acidification of running waters. Acta Botanica Gallica, 142(6), 541–550.

[41] Haury J., Peltre M.C., Muller S., Tremolieres M., Barbe J., Dutartre A., Guerlesquin M. (1996) Des indices macrophytes pour estimer la qualite des cours d'eau francais: Premières proposition. Écologie, 233–244.

[42] Suárez Ma.L., Mellado A., Sánchez-Montoya Ma.M., Vidal-Abarca Ma.R. Proposal of an index of macrophytes (IM) for evaluation of warm ecology of the rivers of the Segura Basin. Limnetica, 24(3–4), 305–318.

[43] Demars B.O.L., Harper D.H. (1998) The aquatic macrophytes of an English lowland river system: assessing response to nutrient enrichment. Hydrobiologia, 384, 75–88.

[44] Caffrey J. (1986) Macrophytes as biological indicators of organic pollution in Irish rivers, in: Richardson D.H.S. (Ed.), Biological Indicators of Pollution. Royal Irish Academy, Dublin, 24–25 february, pp. 77–87.

[45] Schneider S., Melzer A. (2003) The Trophic Index of Macrophytes (TIM) – a new tool for indicating the trophic state of running waters. International Review of Hydrobiology, 88(1), 49–67.

[46] Holmes N.T.H., Boon P.J., Rowell T.A. (1998) A revised classification system for British rivers based on their aquatic plant communities. Aquatic Conservation: Marine Freshwater Ecosystems, 8, 555–578.

[47] Holmes N.T.H., Newman J.R., Chadd S., Rouen K.J., Saint L., Dawson F.H. (1999) Mean Trophic Rank: a users manual R & D Technical Report No. E38. Environment Agency of England & Wales, Bristol, UK.

[48] Newman J.R., Dawson F.H., Holmes N.T.H., Chadd S., Rouen K.J., Sharp L. (1997) Mean Trophic Rank: a user's manual. Environmental Agency of England & Wales, Bristol, UK.

[49] Dawson F.H., Newman J.R., Gravelle M.J., Rouen K.J., Henville P. (1999) Assessment of the trophic status of rivers using macrophytes. Evaluation of the Mean Trophic Rank. Research and Development, Technical Report E39, Environment Agency, Bristol.

[50] AFNOR (2003) Détermination de l'indice biologique macrophytique en rivière (IBMR). AFNOR NF T, 90–395.

[51] Grasmück N., Haury J., Léglize L., Muller S. (1995) Assessment of the bio-indicator capacity of aquatic macrophytes using multivariate analysis. Hydrobiologia, 300/301, 115–122.

[52] Haury J., Muller S. (1991) Variations écologiques et chorologiques de la végétation macrophytique des rivières acides du Massif armoricain et des Vosges du Nord (France). Revue des sciences de l'eau, 4, 463–482.

[53] Zechmeister H.G., Grodzińska K., Szarek-Łukaszewska G. (2003) Bryophytes (Chapter 10), in: Markert B.A., Breure A.M., Zechmeister H.G. (Eds.), Bioindicators & Biomonitors, Principles, Concepts and Applications. Elsevier, Amsterdam, pp. 329–375.

[54] Cenci R.M., Muntau H. (1993) L'utilizzo dei muschi acquatici quali bioindicatori di inquinamento nelle acque da parte di metalli pesanti. Inquinamento, 1, 42–48.

[55] Salemaa M., Derome J., Helmisaari H.-S., Nieminen T., Vanha-Majamaa I. (2004) Element accumulation in boreal bryophytes, lichens and vascular plants exposed to heavy metal and sulphur deposition in Finland. Science of the Total Environment, 324(1–3), 141–160.

[56] Szczepaniak K., Biziuk M. (2003) Aspects of the biomonitoring studies using mosses and lichens as indicators of metal pollution. Environmental Research, 93(3), 221–230.

[57] Figueira R., Ribeiro T. (2005) Transplants of aquatic mosses as biomonitors of metals released by a mine effluent. Environmental Pollution, 136(2), 293–301.

[58] Bleuel C., Wesenberg D., Sutter K., Miersch J., Braha B., Bärlocher F., Krauss G.-J. (2005) The use of the aquatic moss *Fontinalis antipyretica* L. ex Hedw. as a bioindicator for heavy metals: 3. Cd^{2+} accumulation capacities and biochemical stress response of two *Fontinalis* species. Science of the Total Environment 345(1–3), 13–21.

[59] Samecka-Cymerman A., Kolon K., Kempers A.J. (2005) A comparison of native and transplanted *Fontinalis antipyretica* Hedw. as biomonitors of water polluted with heavy metals. Science of the Total Environment, 341(1–3), 97–107.

[60] Rott E. (1991) Methodological aspects and perspectives in the use of periphyton for monitoring and protecting rivers, in: Whitton B.A., Rott E., Friedrich G. (Eds.), Use of Algae for Monitoring Rivers, Institut für Botanik, Universität Innsbruck, Innsbruck, pp. 9–16.

[61] Stevenson R.J., Pan Y. (1999) Assessing environmental conditions in rivers and streams with diatoms, in: Stoermer E.F., Smol J.P. (Eds.), The diatoms: application for the environmental and earth sciences. Cambridge University Press, Cambridge, pp. 11–40.

[62] Descy J.P., Coste M. (1991) A test of methods for assessing water quality based on diatoms. Verhandlungen Internationale Vereinigung für Theoretische und Angewandte Limnologie, 24, 2112–2116.

[63] CEMAGREF (1982) Etude des méthodes biologiques d'appréciation quantitative de la qualité des eaux. Rapport Q.E. Lyon A.F. Bassin Rhône-Méditérannée-Corse, p. 218.

[64] Rumeau A., Coste M. (1988) Initiation à la systématique des diatomées d'eau douce. Bull. Fr. Pêche Piscic., 309, 1–69.

[65] Lenoir A., Coste M. (1996) Development of a practical diatom index of overall water quality applicable to the French national water Board network, in: Whitton B.A., Rott E. (Eds.), Use of Algae for Monitoring Rivers II. Studia Student G.m.b.H, Innsbruck, pp. 29–43.

[66] Leclercq L., Maquet B. (1987) Deux nouveaux indices chimique et diatomique de qualité d'eau courante. Application au Samson et à ses affluents (Bassin de la Meuse Belge). Comparaison avec d'autres indices chimiques, biocénotiques et diatomiques. Institut Royal des Sciences Naturelles de Belgique. Document de Travail, no. 38, Brussels, Belgium.

[67] Schiefele S., Schreiner C. (1991) Use of diatoms for monitoring nutrient enrichment, acidification and impact of salt in rivers in Germany and Austria, in: Whitton B.A., Rott E., Friedrich G. (Eds.), Use of Algae for Monitoring Rivers. Studia Student G.m.b.H., Innsbruck, pp. 103–110.

[68] Sládeček V. (1986) Diatoms as indicators of organic pollution. Acta Hydrochimica et Hydrobiologica, 14, 555–566.

[69] Rott E., Hofmann G., Pall K., Pfister P., Pipp E. (1997) Indikation-slisten für Aufwuchsalgen in österreichischen Fliessgewässern. Teil 1: Saprobielle Indikation. Bundesministerium für Land-und Forstwirtschaft, Wasserwirtschaftskataster, Wien, Austria.

[70] Dell'Uomo A. (1996) Assessment of water quality of an Apennine river as a pilot study for diatom-based monitoring of Italian watercourses, in: Whitton B.A., Rott E. (Eds.), Use of Algae for Monitoring Rivers II. Institut für Botanik, Universität Innsbruck, pp. 65–72.

[71] Dell'Uomo A. (2004) L'indice diatomico di eutrofizzazione/polluzione (EPI-D) nel monitoraggio delle acque correnti. Linee guida. Agenzia per la protezione dell'ambiente e per i servizi tecnici (APAT) - Centro Tematico Nazionale – Acque Interne e Marino costiere, Roma - Firenze, p. 101.

[72] Allan J.D. (1995) Stream Ecology. Structure and Function of Running Waters. Chapman and Hall, New York.

[73] Siligardi M., Cappelletti C., Chierici M., Ciutti F., Egaddi F., Maiolini B., Mancini L., Monauni C., Minciardi M.R., Rossi G.L., Sansoni G., Spaggiari R., Zanetti M. (2003) Indice di Funzionalità Fluviale I.F.F. Manuale di applicazione. Manuale ANPA/2° Edizione.

[74] Vannote R.L., Minshall G.W., Cummins K.W., Sedell J.R., Cushing C.E. (1980) The river continuum concept. Canadian Journal of Fisheries and Aquatic Sciences, 37(1), 130–137.

[75] Siligardi M., Maiolini B. (1992) L'inventario delle caratteristiche ambientali dei corsi d'acqua alpini: guida all'uso della scheda RCE-2. Biologia Ambientale, VII(30), 18–24.

[76] Ormerod S.J., Wade K.R., Gee A.S. (1987) Macro-floral assemblages in upland Welsh streams in relation to acidity, and their importance to invertebrates. Freshwater Biology, 18(3), 545–557.

[77] Monteith D.T., Evans C.D. (2005) The United Kingdom Acid Waters Monitoring Network: a review of the first 15 years and introduction to the special issue. Environmental Pollution, 137(1), 3–13.

[78] Kilgour B.W., Barton D.R. (1999) Associations between stream fish and benthos across environmental gradients in southern Ontario, Canada. Freshwater Biology, 41, 553–566.

[79] Paavola R., Muotka T., Virtanen R., Heino J., Kreivi P. (2003) Are biological classifications of headwater streams concordant across multiple taxonomic groups? Freshwater Biology, 48(10), 1912–1923.

[80] Heino J., Parviainen J., Paavola R., Jehle M., Louhi P., Muotka T. (2005) Characterizing macroinvertebrate assemblage structure in relation to stream size and tributary position. Hydrobiologia, 539(1), 121–130.

4 Marine organisms as biomonitors

M E Conti & M Iacobucci

1 Introduction

Biomonitors are organisms mostly used for the quantitative determination of contaminants in the environment (see Chapter 1). The present chapter discusses mainly the employment of seaweeds, sea phanerogams and molluscs as bioindicators of heavy metal contamination, since they are the organisms of choice in most of the programmes of biological monitoring of coastal marine waters (see Water Framework Directive 2000/60 EC (WFD) [1]).

Bioindicators provide an integrated view of the presence of micropollutants, and in the aquatic environment they respond only to the fraction presenting a clear ecotoxicological relevance (see Chapter 1).

Among the different methodological approaches of biological monitoring of a marine ecosystem, the approach based on the bioaccumulation capacity of some chemical species, as for instance trace metals, is among the most important. This capacity is particularly present among macroalgae, sea phanerogams, molluscs, oysters, etc. These organisms can be proposed as possible bioindicators when the presence of a direct correlation between internal and environmental levels of the various metals has been verified [2].

Therefore, the chemical analysis of the tissues of such organisms represents a method for measuring the integrated bioavailability of trace metals in the marine environment over time, given their remarkable bioconcentration capacity [3–5].

Comparing the levels of metals in the organisms collected at different sites is a very useful way to gather information on the biologically available levels of inorganic contaminants in the areas under study. In general, micro and macroalgae, marine phanerogams, mussels, oysters, tellinid bivalves, polychaets and crustaceans are some of the organisms used as bioindicators of marine ecosystems [4].

Marine organisms can accumulate contaminants present in solution and in the particulate material, thus determining a variation in the level of bioaccumulation depending on the trophic level occupied. Some marine invertebrates can take contaminants from the particulate material of the sediment, especially when their physico-chemical conditions are being altered. Algae and phanerogams take contaminants in solutions and many herbivorous crustaceans reflect their concentrations.

Therefore, in order to have a full picture of the bioavailability of the various pollutants of water ecosystems, it is necessary to use several kinds of bioindicators: macrophytes algae responding mainly to metals in solution, marine phanerogams, which are higher plants provided with a root apparatus and responding also to the amount of metals biologically available in sediments; suspensivorous filter feeders (like mussels) responding to dissolved and suspended contaminants and detritivores taking recent deposits on the sediment.

2   Algae and bioaccumulation of heavy metals

Heavy metals are natural constituents largely distributed in the environment; when highly concentrated they are harmful to animals and plants. Moreover, they cannot be eliminated from the water body: they settle on the ocean bottom, and currents bring about a slow release which causes serious risks of contamination for the organisms [6].

The existing balance between suspended metals and metals present on the sediments was in many cases broken by human activity which caused a wholesale increase in the concentration levels of metals through uncontrolled discharges, especially in highly industrialized areas. In chronically polluted areas, algae and sea plants tend to accumulate metals in very high and dangerous levels [7, 8].

The levels of bioaccumulation of contaminants in older algal tissues can be compared to the accumulation on more recent tissues of the same individuals and thus help determine the evolution of the contamination process for the organisms present in a given geographical area.

Sea macroalgae belong to divisions *Chlorophyta, Rhodophyta and Phaeophyta.* From *Chlorophyta* the species of green algae (*Chlorophyceae*) pertaining to the genera *Enteromorpha* [6, 9–16], *Ulva* [11–13, 17–25], *Cladophora* [10, 12, 26], *Codium* [11, 27], *Caulerpa* [11, 12] and *Chaetomorpha* [10] have been selected as bioindicators of heavy metals in the marine environment.

From *Phaeophyta* the species of brown algae (*Phaeophyceae*) used in monitoring studies belong to the genera *Dictyota* [11, 28], *Scytosiphon* [28], *Colpomenia* [29, 30], *Padina* [16, 31–33], *Fucus* [34–37], *Ascophyllum* [38, 39] and *Macrocystis* [29, 40].

From *Rhodophyta* the species mostly used were the red algae (*Rhodophyceae*) of the genera *Gracilaria* [10, 12, 17–19], *Pterocladia* [11, 28], *Gelidium* [28], *Gigartina* [11, 29, 30] and *Porphyra* [28].

Not only the selection of the appropriate bioindicators but also the knowledge of biological variables (age, sex, size, seasonality, portion of the plant analysed) related to the specific organism selected or particular features of the algal species (degree of submergence in the sea, tolerance to metal stress) may be taken into account in the sampling design [41, 42].

In algae, apart from sampling considerations (see below), metal concentrations depend both on external factors (pH, salinity, inorganic and organic compounds) and on physico-chemical parameters (temperature, light, oxygen and nutrient) [8]. Metal bioaccumulation is a function of the total enrichment in metal by a specific algal organism, taking into account those metal forms, known as bioavailable species, that are more easily integrated by the organism [43].

Sea life benefits from algae, which are at the basis of the food chain. The net amount of global oxygen produced by algae through photosynthesis is an estimated 30–50%. Macroalgae have a multicellular organization, their structures can be complex and quite large in terms of size, and sometimes they have a differentiated thallus in organs analogous to those of higher plants [44]. Some chlorophyte like *Caulerpa* and others are multinucleated unicellular organisms equally able to show complex structures [45].

The degree of metal bioaccumulation in aquatic organisms has been widely studied and can serve as a reference point for further monitoring studies as well as to help prevent potentially harmful effects on higher organisms.

The earliest studies, dating back to the 1960s [46, 47] found a linear correlation between zinc concentrations in *Ulva* and *Laminaria digitata* and mean concentrations of zinc in the surrounding water. Also Seeliger and Edwards [48] showed an existing correlation of 0.98 for copper and 0.97 for lead between seawater and a number of algal species (*Blidingia, Enteromorpha and Fucus*). Generally, suspended metals present in water are a function of the kind and quantity of sediments and only partially do they depend on climatic and hydrographic processes [8, 48].

Biomonitoring studies require therefore rigorous sampling procedures. Moreover, metal concentration in algae can vary along both the vertical and the horizontal position of the water column, in relation to the variation of the degree of exposure to contaminants [2, 47].

Algae only bind free metal ions whose concentrations depend in turn on the nature of suspended particulate matter which is formed by organic and inorganic compounds [8, 49]. At an early stage, algae accumulate metals on their external surface through a reversible physico-chemical uptake process, followed by a process regulated by cellular metabolism [8, 50]. This kind of uptake has been the object of debate among researchers. Other authors, for instance, found that zinc accumulation in some brown algae is irreversible [35, 51].

In general, it is widely accepted that the metal uptake through cellular membranes is mainly due to diffusion rather than an active transport mechanism.

The kinetics of uptake through passive diffusion is described by Fick's law. Depledge and Rainbow [52] suggest that Fick's equation may require some modification in order to take into account also the differences in the electrical potential existing through many cellular barriers [52].

Figure 1 shows a general model of the distribution of metals in aquatic ecosystems.

The accumulation mechanisms of metal on the part of algae takes place through the way metals bind to the external cellular wall. This can be of an ionic kind or through the formation of complexes with the ligands of the cellular wall. The polymers constituting the cellular wall are rich in carboxylic, phosphoryl, aromatic and hydroxyl groups that can bind cationic metals [53, 54].

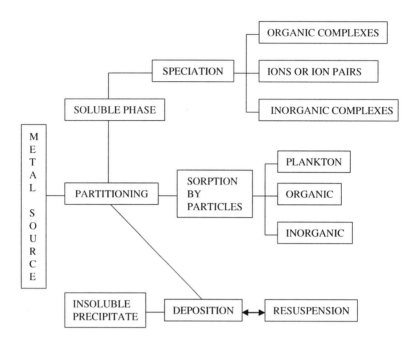

Figure 1: General model of the distribution of metals in an aquatic ecosystem.

They can also produce natural organic complexes which can affect the uptake of metals to varying degrees. Several studies demonstrate that algae reduce the toxicity of Cu through a reduction of ionic activity [55, 56]. In particular, copper and mercury have a strong tendency to form organic complexes when compared to other heavy metals, even though both coastal and estuary organic complexes contain several ligands presenting different stability constants.

Macroalgae live in a saline environment and therefore require ion exchange mechanisms to selectively absorb essential metal ions such as calcium and potassium. Among the other functions of these polysaccharides in gel form present on the external cellular walls of algae there may be the defence of other organisms and the increase in elasticity of the cellular wall and consequently a better ability to adapt of the organism to the wave motion [57]. For instance, the accumulation of strontium is regulated by the selectivity of algal polysaccharides.

At high concentrations, all heavy metals are toxic to algae. Although, in absence of essential metals (Zn, Cu, Mo, etc.) algae stop growing effectively because physiological and/or biochemical processes are affected. Heavy metals tend to inhibit algal photosynthesis [58]. Cu is an essential element for algae, and its toxic effects have been widely studied [59–61]. At high concentrations, Cu inhibits photosynthesis. Its toxic action affects the membrane, causing the loss of K^+ ions and consequently brings about variations in the cellular volume also involving chloroplasts, thus inhibiting the electron transportation [59, 62].

Cadmium is one of the most toxic heavy metals and is considered as non-essential for living organisms. This metal competes for Zn sites when present in high concentrations in the water body and it can cause considerable morphological, biochemical and physiological alterations in algae, protozoa and cyanobacteria [63]. Cadmium causes a decrease in cell volume, growth rate and levels of photosynthetic pigments [64]. Devi Prasad and Devi Prasad [65] studied the effects of 0.1–10 ppm of Cd on three families of green algae and found that at a concentration of 5 and 10 ppm there is a total inhibition of the growth process. Algae are also very sensitive to the presence of nickel in the water body. Devi Prasad and Devi Prasad [65] observed tolerance levels of 10 ppm of Ni for green algae, while Çalgan and Yenigün [66] detected a total inhibition of growth and nitrogenase activity with a Ni concentration of 5 ppm for cyanobacterium *Anabaena cylindrica*.

The knowledge of the degree of tolerance of algal species to some heavy metals over time is another important factor to take into account in monitoring studies. Tolerance ranges to metals can be established, for instance, by adding (generally in laboratory studies) heavy metals over a given period of time and then performing a control over the photosynthetic activities of the algae under study.

It is generally accepted that there is a correlation between the tolerance of algae to heavy metals and metal concentrations in the environment where algae develop [67, 68]. Probably, over the last 100 years there have been selection phenomena of tolerance to heavy metals, particularly in sites close to foundries or mines. More studies are needed in order to express a valid conclusion on whether the various degrees of tolerance to heavy metals are genetically steady or not. However, tolerance to Cu of eukaryotic algae seems to be genetically steady for a period of at least two years [69]. The attainment of algal families with a high degree of tolerance to heavy metals could also be the result of laboratory selection phenomena of spontaneous during the culture. Moreover, the presence of plasmids has been postulated: they are thought to control the tolerance of cyanobacteria to Cu [69].

Algae can develop detoxification mechanisms. The production of phytochelatins is the generally accepted system for the detoxification of metals. It has been proven that Cd is the strongest phytochelatin synthesis inductor. Also Cu and Zn have the same effect, although not so strong as Cd [70, 71].

As to growth, metal concentrations in macroalgae are observed on the tips (which reflect the most recent pollution, given their typical rapid growth) and on the thallus, which is considered the integrator of the contamination processes in the life of the organism. Metal concentration increases with age and is therefore higher on the thallus than it is in tips [72].

When concentrations in the water body are particularly high, the various metals can present interference phenomena (synergy and antagonism effects). Metal concentrations in algae depend both on external factors, involving the interaction between metal and cell wall (pH, salinity, inorganic and organic compounds) and on physico-chemical parameters, such as temperature, light, oxygen and nutrients [2, 73].

The bioavailability of a pollutant determines its true ecotoxicological value. Its chemical form (speciation-oxidation status) and its reactivity are of highly relevance than its accumulation level in the water body.

Also temperature and salinity play an important role. Seawater salinity is in average 35 g/kg_{sw} (sw = seawater), presenting values ranging between 33–37 g/kg_{sw} in the offing. These values can be higher in isolated basins and can be lower where there is a considerable flow of fresh water. For instance, *Ulva rigida* accumulate large amounts of Fe, Zn and Cd when salinity is low.

Generally, macroalgae are used in bioaccumulation tests, while eukaryotic and prokaryotic microalgae are more extensively used for ecotoxicological tests [74]. Ecotoxicological tests on animals and plants (i.e. microalgae *Dunaliella terti-olecta*) are usual in environmental monitoring along with the analytical determinations of several organic and inorganic pollutants. Such tests are performed at different trophic levels so as to give a global assessment with the potential risks associated with the studied areas [75].

Among the most studied microalgae there is the species *Chlorella*, which has remarkable bioaccumulation capacities of a wide range of metal cations such as for instance copper and nickel [76, 77], cadmium and zinc [77, 78], lead [79] and rare earth elements (REEs) [80]. This species requires also a minimum nutritional input and this makes the use of these algae much more convenient as they turn out to be a valid metal decontamination method of waste waters [81, 82]. *Scenedesmus, Selenastrum* and *Chlorella* (fig. 2) have a high capacity of accumulating metals (between 6 and 98%). In general the uptake of metal ions on the part of *Chlorella regularis* depends more on the physico-chemical uptake at the level of the cell wall than on its biological activities. For *Chlorella regularis* the selectivity of metal ion uptake is mainly due to the strength of the bond between metal ion and cellular components, particularly proteins (metallothioneins (MTs)). These have an important role for the accumulation of ions of heavy metals on the part of microorganisms (see Chapter 1) [83].

In the Mediterranean Sea there are approximately 130 species of green algae (*Chlorophyceae*). They reach the peak of their growing period between April and August, are very strong and are easily adaptable to stagnation, warming of the

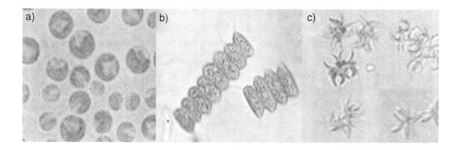

Figure 2: (a) *Clorella spp.*, (b) *Scenedesmus spp.*, (c) *Selenastrum spp.*

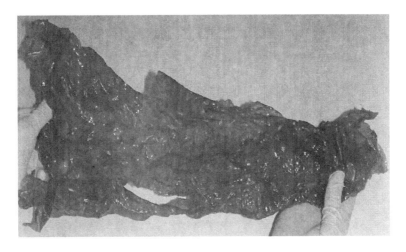

Figure 3: Sample of *Ulva rigida* C. Ag. collected at the Beagle Channel, Tierra del Fuego, Argentina.

water body and low amount of oxygen, even when waste waters are produced. The genus *Ulva* is probably the most commonly used in biomonitoring studies of the Mediterranean Sea; it is also found in the Beagle Channel (Patagonia, Argentina) (fig. 3) where it thrives and can be as long as one metre along the coastline, whereas in exposed areas it is only a few centimetres long. The species *Ulva lactuca* (L.) is especially present in polluted areas close to the surface – never deeper than 4–5 m – and it can be as long as 40 cm.

Table 1 shows literature data regarding the bioaccumulation of heavy metals for *Ulva* in several geographical sites.

As above reported, algae of the genus *Enteromorpha* can be used as bioindicators in the Mediterranean Sea; they are largely widespread in shallow waters and puddles left by the ebbing tide and are often associated to the genus *Ulva*. They are found in eutrophic and polluted waters. Species typical of the Mediterranean Sea are *Enteromorpha intestinalis* (L.) Link and *Enteromorpha linza* (L.) J. Ag. [6, 9–16].

Other species of green algae used as bioindicators, although less often, are seaweeds belonging to the genera *Cladophora* and *Chaetomorpha*. *Chaetomorpha aerea* (Dilw.) Kütz. is very commonly found on rocks, in shallow pools left by the ebbing tide, especially in spring and autumn. *Chaetomorpha capillaris* (Kütz.) Borg. is widespread on coastlines, detached from the substrate and associated to larger algae [10].

The Genus *Codium* [11, 27] is used as bioindicator, although less commonly than other species. There are several species that can be found as deep as 50 m, such as *Codium bursa*, common on sandy seabed. Costantini *et al.* [27] detected very moderate levels of mercury for *Codium tomentosum*: out of 11 sampling sites all over Italy only two samples yielded noticeable results (8 and 10 ppb fresh weight).

Table 1: Selected references of trace metal concentrations (μg/g d.w.) in *Ulva spp.* from various geographical locations.

Species	Sites	As	Cd	Cr	Cu	Fe	Ni	Pb	Zn	References
Ulva rigida	Venice Lagoon, Italy		0.1–0.6	0.1–14	2.11–21			0.5–7.01	8–110	Favero *et al.* [23]
Ulva lactuca	5 sites in Gaeta-Formia Gulf, Italy		0,12 0.035–0.20	1,36 0.34–2.4				2,05 0.51–3.12		Conti [84]
Ulva rigida	7 sites in Thermaikos Gulf, Greece		1,0 0.1–2.5		2,2 1.1–4.3	$0,08–0,12 \times 10^3$		14,7 6.3–29.8	57,3 39.0–82.5	Haritonidis and Malea [85]
Ulva lactuca	Gulf San Jorge, Argentina		0.12	1.56	5.5			3.6	5.2	Muse *et al.* [13]
Ulva lactuca	Province of Quebec, Canada	6	0.22	4.4	19.2	2486		1.64	33.3	Phaneuf *et al.* [86]
Ulva spp.	Aegean Sea, Greece		0,24–1,1		7–14,5			0,02–2,8	16,4–88	Sawidis *et al.* [87]
Ulva spp.	Rias Baixas, Spain				2,18–20,6	$0,039–2,62 \times 10^3$			6,96–66,8	Villares *et al.* [88]
Ulva lactuca	Apulian coast, South Adriatic Sea, Italy	0.2		12.07	337.1			0.84	127.27	Storelli *et al.* [89]
Ulva rigida	Venice Lagoon, Italy	7	0.2	4.6	13	1033	2.6	7.3	64	Caliceti *et al.* [90]
Ulva lactuca	Gulf of Suez, Red Sea		0,69–1,48		8,88–13,27			3,87–23,29	15,23–49,65	El-Moselhy and Gabal [91]
Ulva lactuca	El Jadida city, Morocco		0,7–1,07		11,9–14,5				163–228	Kaimoussi *et al.* [92]

The majority of brown algae (*Phaeophyceae*) live in shallow waters, only a few metres deep. Among the brown algae (see also above) also important are the species of the orders *Laminariales* and *Fucales* [34–37] largely used in biomonitoring studies for other seas apart from the Mediterranean. Among the *Fucales* used in biomonitoring studies, there is the genus *Cystoseira* (fig. 4). The species of this genus are largely diffused along the western coasts of the Adriatic Sea. Brown algae have a high selectivity for divalent metals: they are able to extract them from the seawater with subsequent release of monovalent ions or magnesium ions [93]. Several studies (not specifically relating to the Mediterranean) show the high capacity of brown algae to accumulate zinc in tissues [35, 51, 94]. *In situ* studies focusing on the transplant of brown algae from a contaminated site to a "clean" one prove that heavy metals are released in the water body to a significant degree [94]. However, the release and the uptake of zinc and cadmium seem to require an input of metabolic energy while the accumulation of lead and strontium is regulated by a process of ionic exchange [93].

In the Mediterranean, approximately 350 species of red algae have been described. They grow well at different depths, but especially between 30 and 60 m where they outnumber brown or green algae. They are generally annual species. As above reported, among the red algae most commonly used in bio-monitoring studies in the Mediterranean are the species of the genera *Gracilaria* [10, 12, 17–19], *Pterocladia* [11, 28] and *Gelidium* [28]. The bioaccumulation of Fe, Cu and Co over the seasons is more relevant for *Ulva lactuca* than it is for the red alga *Gracilaria verrucosa* [95]. This is because *Ulva lactuca* has higher protein concentration than *Gracilaria verrucosa*, which has high quantities of

Figure 4: *Cystoseira spp.* (Ustica island, Sicily, Italy).

carbohydrates. As a matter of fact, *Ulva lactuca* needs nutrient salts rich in nitrogen and therefore it is mostly found in biotopes close to areas with urban or industrial waste waters, generally biotopes with a high level of heavy metal pollution load [95].

Generally, heavy metal bioaccumulation studies are numerous and involve many algal species. However, the speed of accumulation or the correlation between metal concentration in seawater and in the algal tissues is still not well known, and further studies are necessary in order to clarify metal uptake mechanisms. In the Mediterranean Sea, green macroalgae of the genus *Ulva* are more numerous and contribute substantially to the biomass input and sensibly to the cycle of mineral elements in coastal ecosystems. These algae are also used as possible eutrophication indicators [96].

3　Phanerogams

Phanerogams have a high capacity to accumulate metals in proportion to the normal levels of concentration in the water environment [2, 6, 97]. Among the most studied kinds of phanerogams we find the species of the genera *Posidonia* and *Zostera* [98]. Phanerogams are plants with a trunk, leaves, roots and rhizomes. Therefore, they happen to provide a greater amount of information about the environmental metal content than algae do, due to the fact that their rhizomes and roots are able to interact with sediments.

In the Mediterranean Sea, for instance, the most studied species is the *Posidonia oceanica*, which is endemic and has been widely studied both *in situ* and in laboratory [2, 6, 20, 27, 97, 99–106].

Posidonia meadows cover vast regions of the submerged areas down to 40 m depth and they also constitute an ideal habitat for several species of animals some of which belong to our food chain. Many national and international programmes deal with the safeguard of the *Posidonia* ecosystem, as it is threatened by anthropic impact and by competition with the tropical alga *Caulerpa taxifolia*, which was accidentally brought in the Mediterranean in 1984 [2, 107–109].

The *Posidonia* leaves have a basal kind of growth; the tip is the oldest part of the leaves and therefore it has a greater time of exposure in the water environment and a higher accumulation of metals. The concentrations of lead, cadmium and chromium in the *Posidonia* leaves follow the sequence: leaf tips > whole leaves > leaf basal tissues [6] (see fig. 6). From several *in situ* tests we can see that cadmium and zinc mainly accumulate in the leaf tissue rather than in rhizomes [6, 97]. This might depend on the different metal distribution between soluble and insoluble (particulate and sediment) fractions. Generally, heavy metal concentrations in the *Posidonia* leaves follow the sequence Zn > Cu > Cd > Pb > Cr. This is the bioaccumulation sequence normally reported in the literature. It is possible to have occasional variations in this pattern, such as the inversion of Cd and Pb [101, 102, 110, 111]. It has been proven that in the *Posidonia* leaves the concentrations of Cd, Cr, Pb and Zn decrease in the

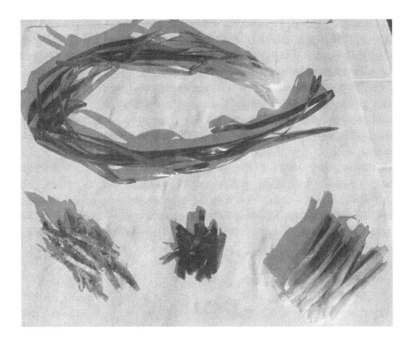

Figure 5: Whole leaves, basal leaves, tips leaves and rhizomes of *Posidonia oceanica* (Linosa island, Italy).

following order: leaf tip > whole leaf > leaf basal tissue (see fig. 5). This trend is in agreement with the observation that the seagrass leaves grow at the basal level. Some authors have used *Posidonia oceanica* as a possible bioindicator for the presence of mercury in the Italian sea, detecting concentrations in the range of 5–38 ppb wet weight [27].

Heavy metal concentrations in the *Posidonia* leaves are greater than in those of *Halophila stipulacea* (a marine phanerogam of the Indian Ocean) or *Zostera marina* (a phanerogam still not very common in the Mediterranean). In the leaves of *Zostera marina* heavy metal concentration follows the sequence: Zn > Cu > Pb > Cd. In general, the variability of metal content in the seagrasses is quite high [110, 112–114]. A study reports that the accumulation of copper and nickel is smaller in *Posidonia australis* [115] than it is in *Posidonia oceanica*.

The high variability detected in the metal concentration of phanerogams depends on the pre-treatment stage of samples, especially the mineralization process, interference phenomena, the matrix effect, standard solutions and the different instruments of analysis [98, 116]. In particular, the sample pre-treatment plays a fundamental role in the unwanted variations of the results. For instance, the kind of rinsing procedure (with seawater or distilled water) can decrease the level of many metals (Cd, Cr, Fe, Mg, Na, Pb, Ti, Zn) present in the *Posidonia* leaves that are being treated [117]. The removal of epiphytes is also very important as they are able to actively accumulate Pb from the water column and consequently alter the actual concentrations of the leaves [118]. Rinsing epiphytes can also cause damages to the tissues and, as a result, an alteration of the final results, depending on the technique adopted. Another important factor, generally overlooked, is the assessment of the organisms' dry weight. According to our experience [2, 6, 97, 119–122] there is a high variability when data are obtained and reported for fresh weight. The assessment of the exact dry weight is associated to fewer mistakes and contrary to wet weight it is generally steadier in the organism.

Other sources of variability often quoted in the literature, and which must be carefully considered, are the mechanisms linked to the physiology of the seagrass species, for instance, kind of tissue, age, growth, accumulation, desorption; as well as physico-chemical conditions of the environment (pH, time of exposure, contamination levels, salinity, etc.) [2]. Figure 6 shows a whole plant of *Posidonia*.

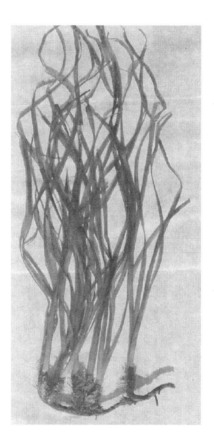

Figure 6: *Posidonia oceanica* collected at the Linosa island (Italy).

4   Invertebrate organisms

4.1   Mussels

Among the mussels generally used in studies on the impact of heavy metals and other contaminants, *Mytilus edulis* and *M. galloprovincialis* in temperate waters, and *Perna* in tropical waters can be mentioned [2, 6]. These molluscs filter the surrounding water and, in so doing, they constantly accumulate metals in their tissues. Also other *Mytilus* species can be employed in these kinds of studies such as *M. chiloensis* (fig. 7) used for biomonitor the Beagle Channel coasts in Tierra del Fuego, Argentina [123].

In different molluscs, crustaceans and fish some metals do enter by active transport. For instance, Cd can be absorbed via active transport through Ca ion pumps. Also the pinocytosis process is involved in the uptake of metal-rich particles in the gills of some molluscs [124, 125]. Ions entering the organism through this process must nonetheless be available, although they might not be absorbed by the organism if these biological processes requiring energy do not occur. Inside the organism metals can associate with ligands with a strong binding capacity.

Generally, in animals we find the occurrence of MTs, cysteine-rich proteins (see also Chapter 2), which strongly bind metals like Cu, Cd and Zn. These proteins are present both in a number of invertebrates and vertebrates and are

Figure 7: *Mytilus chiloensis* H. collected at the Beagle Channel, Tierra del Fuego, Argentina [123].

inducible by exposure to raised environmental concentrations of some heavy metals [124]. In higher plants, fungi and algal cells there is the presence of a family of polypeptides called phytochelatins, which seem to have an important role in the homoeostatic control and the sequestration of metal ions [126]. Both MTs – in some animal organisms – and phytochelatins in phytoplankton could be used as quantitative markers of metal exposure in the marine environment. As to the use of these biomarkers which characterize metal exposure, it is necessary to previously and accurately assess the normal metallothionein and phytochelatin concentrations in the tissues under study [127, 128].

The mussel is a partial regulator of Cu and Zn but it is also a useful bioindicator of many other heavy metals. The biochemical processes of metals inside the cells represent the main factor regulating the concentration and the distribution of metals in molluscs. The best instance of a ubiquitous metal sequestration system is that of MTs. As previously reported, MTs serve as a buffering system for the intracellular metal ions and they are found in many branches of the phylogenetic tree [2, 6].

Metal accumulation can either follow a linear or an exponential function. In general, early in the bioaccumulation process there is an exponential function when the uptake is fast in proportion to the release. The linear accumulation over time increases when the metal excretion is negligible or very slow in relation to the uptake, so the concentrations in the organism rise during the time of exposure. Hence, sites that bind metals do not become saturated, or the organism's growth rate does not exceed the metal uptake. This bioaccumulation pattern is quite common, for instance, in molluscs exposed to Cd, and in many cases it is triggered on by the increasingly irreversible link with the buffering systems (e.g. MTs). In general, if the involvement of MTs in the metabolism of a metal is limited – and its excretion is therefore no longer unimportant – we are in the presence of an exponential model [124].

Metal concentrations in the organisms are a function of the net accumulation determined on their weight. This means that the animal's dimensions affect the metal concentration in the organisms to a remarkable degree. Some laboratory studies on bivalves, crustaceans and some kinds of fish, reveal that a faster accumulation occurs in small size individuals, as they have a higher area/volume ratio. On the other hand, *in situ* tests show that the effects deriving from the dimensions of the organisms vary with the species, the metal and, quite often the site examined [2]. The relation between the metal concentration (Y) and the weight of the organism (W) is known as power function [129, 130]. The relation between these parameters can be turned into a linear function through a logarithmic transformation expressed in the following equation:

$$\log Y = \log a + b \log W$$

where a is the multiplicative coefficient of the power function and b is the gradient of the linear function thus obtained, which coincides with the power function exponent [131].

If the slope (b) in the equation is $b = 1$, the metal accumulation and the metabolism take place at a rate that is proportional to the growth and is therefore linked to the metabolism of the animal. If the slope is <1, the smaller the animals the faster the accumulation, maybe because of the higher area/volume ratio. A slope >1 means a constant accumulation throughout the organism's lifetime, never reaching, for this reason, a steady state. The younger the animals the faster the growth; as a consequence, the enlargement of the tissues causes a dilution in metal concentrations, which tend to decrease with the ageing of the individual. Boyden, in his studies on several species (bivalve and gastropods molluscs, oysters, etc.), has found that the majority of the slopes for Cd, Co, Cu, Fe, Mn, Ni, Pb and Zn are quite close to 1 (metal content is independent of the dimension) or <1 (metal contents are higher in smaller individuals) [2, 6].

Therefore, the variability compensation in metal concentrations linked to the organisms' growth must be an important aspect in the elaboration of biomonitoring programmes [130, 132]. Although mussels are greatly responsible for the theoretical production concerning the bioaccumulation of metals, other gastropod molluscs (e.g. *Patella spp.*; *Monodonta spp.*; etc.) are suggested as possible biological indicators for the presence of trace metals in the aquatic ecosystems [97, 122, 133]. In some cases the shells of these organisms have been tested as possible bioindicators of heavy metal in the environment but in a test carried out on the shells of *Patella vulgata* the hypothesis of their use in environmental trace monitoring was rejected because they do not reflect the environmental availability of metals [134]. It must be noticed that the importance of the metal source (water or food) in the molluscs depends on the eating habits of the animal. Metals in solution generally have a higher bioavailability with respect to solid phase metals (i.e. those metal species assimilated through the diet); however, since the concentrations of metals in solids are higher they have a greater importance as contamination carriers [2].

In biomonitoring studies it is important to consider also the source of food of the animal under study. For the gastropod molluscs quoted as *Monodonta* and *Patella*, which are herbivorous, we can assume a possible correlation with the metal contents of the algae which are their staple food. The same assumption can be extended to seagrasses which might reflect the metal concentrations entering through the root from the sediments.

Monodonta turbinata and *Patella caerulea* have been used to classify uncontaminated ecosystems in southern Tyrrhenian areas [97, 135–138]. These species, particularly *Monodonta*, showed very good levels of metals bioaccumulation. With the appropriate statistical treatment of data (see Chapter 7) it is possible to compare metal pollution levels of different marine ecosystems [139]. Figure 8 shows individuals of *Monodonta* and *Patella*.

This approach is of paramount importance because it allows to have relevant information on basal pollution levels of ecosystems of different geographical areas. This also allows to have reliable information for the formulation of environmental programmes.

Figure 8: *Monodonta turbinata* (left) and *Patella caerulea* (right) individuals collected at Linosa island (Italy).

Many international projects have been conducted on this matter. For instance, the Mussel Watch Programme (see web site: www.ccma.nos.noaa.gov) analyses chemical and biological contaminant trends in sediment and bivalve tissues collected from about 250 coastal and estuarine sites from 1986 to present in USA.

4.2 Detritivores

Detritivores are scavengers which feed on dead plants and animals or their waste; they concern mainly those species of tellinid bivalves (*Scrobicularia plana, Macoma balthica*) feeding on the recent deposit on the sediment; they are good bioindicators but have a limited distribution in comparison with *Mytilus*.

Several species of polychaets are used as bioindicators. *Nereis diversicolor* can regulate some metals like Mn and Zn, but it is however widely used because it is largely distributed.

Some decapod crustaceans (crayfish, crab, etc.) have a wide distribution but their mobility and regulation capacity can create some problems in the interpretation of analytical data. Among the amphipods there are several species (*Orchestia gammarellus, O. tenuis, Platorchestia platensis*) that have a potential value as bioindicators. Also Cirripedia are excellent bioindicators of trace metals and are well distributed in the environment.

Table 2 reports selected references of metal concentrations in marine organisms.

Table 2: Selected references of trace metal concentrations (µg/g d.w.) in *P. oceanica, P. pavonica, Cystoseira spp., M. turbinata* and *P. caerulea* from various geographical locations.

Species	Sites	Cd	Cr	Cu	Pb	Zn	Notes	References
Posidonia oceanica	Ischia Island, Italy	1		14.1	3.4	168	Leaves	Schlacher-Hoenlinger and Schlacher [140]
	Arburese Coast, Sardinia, Italy	0.6		27.1	5.2	109.2	Rhizomes	Caredda et al. [141]
	El Portús, Spain	1.0–8.6			0.7–10	105–180	Leaves	Sanchiz et al. [142]
	Favignana Island, Sicily, Italy	2.22	0.50	11.6	0.91	112	Leaves	Campanella et al. [97]
	Middle Adriatic Sea	0.35–2.01			0.61–6.82		Leaves	Kljaković-Gašpić et al. [143]
	Islands Egadi, Sicily, Italy	2.7–3.1		7.0–13	8.8–14	190–220	Leaves	Tranchina et al. [144]
	Trapani, Sicily, Italy	2.3–3.1		8.0–11	7.4–13	170–280	Leaves	
	Marsala, Sicily, Italy	1.9–3.1		2.5–6.2	7.5–9	290–700	Leaves	
	Carini, Sicily, Italy	3.3–4.0		10.0–12	9.8–12	180	Leaves	
	Ustica, Sicily, Italy	5.98	0.35	31.88	2.29	213.13	Leaves	Conti et al. [136]
Padina gymnospora	Praia Grande, Septiba Bay, Rio de Janeiro, Brazil	0.61–1.20				101–610	Collection dates: from 04/90 to 07/97	Amado Filho et al. [145]

(Continued)

Table 2: (Continued)

Species	Sites	Cd	Cr	Cu	Pb	Zn	Notes	References
	Ilha do Gato, Septiba Bay, Rio de Janeiro, Brazil	0.98–2.70				166–899	Collection dates: from 04/90 to 07/97	
Padina durvillaei	Loreto Bay, Baja California, Mexico		2.55–2.77			36–83		Sánchez-Rodríguez et al. [146]
Padina pavonica	Favignana Island, Sicily, Italy	1	2.86	11	6.36	53		Campanella et al. [97]
	Gulf of Gaeta, Italy	0.5	3.45	12.3	3.98	51		Conti and Cecchetti [135]
	Ustica, Sicily, Italy	0.71	0.91	6.83	3.89	49.61		Conti et al. [136]
Cystoseira barbata	Venice lagoon, Italy	0.1	1.1	7	3.2	38		Caliceti et al. [90]
	Şile (Black Sea)	0.78	< 0.06	3.43	1.4	21.7	2000	Topcuoğlu et al. [147]
	Sinop (Black Sea)	0.09	1.2	1.7	3.5	6.5	2000	
Cystoseira spp.	Ustica, Sicily, Italy	0.28	0.54	11.27	5.24	144.84		Conti et al. [136]

Patella caerulea	Favignana Island, Sicily, Italy	4.41	0.3	1.7	0.2	5		Campanella *et al.* [97]
	Gulf of Gaeta, Italy	3.54	0.85	14.3	0.95	100.8		Conti and Cecchetti [135]
	Iskenderun Bay, Northern East Mediterranean Sea	2.39–4.97	4.77–8.33	1.58–4.02	4.28–14.5	23.1–46.6		Türkmen *et al.* [148]
	Ustica, Sicily, Italy	4.7	0.56	6.27	1.02	51.39		Conti *et al.* [136]
Monodonta mutabilis	Nickel Smelter, Greece	3	9	222.8		88.6	Viscera polluted	Nicolaidou and Nott [149]
		3.6		43.15		109	Muscle polluted	
		5.8	25.2	222.6		135.3	Viscera clean	
M. labio	Cape d'Aguilar Marine Reserve, Hong Kong	5.8		26.4		55.9		Blackmore [150]
M. turbinata	Favignana Island, Sicily, Italy	1.44	0.31	10.9	0.24	31		Campanella *et al.* [97]
	Gulf of Gaeta, Central Italy	1.12	0.42	62.2	0.58	98		Conti and Cecchetti [135]
	Ustica, Sicily, Italy	1.66	0.68	22.12	0.69	64.01		Conti *et al.* [136]

References

[1] Directive 2000/60/EC of the European Parliament and of the Council of 23 October 2000 establishing a framework for Community action in the field of water policy. Official Journal L. 327 of 22.12.2000.

[2] Conti M.E. (2002) Il monitoraggio Biologico Della Qualità Ambientale, SEAM, Roma, p. 180.

[3] Phillips D.J.H., Segar D.A. (1986) Use of bio-indicators in monitoring conservative contaminants: programme design imperatives. Marine Pollution Bulletin, 17, 10–17.

[4] Rainbow P.S., Phillips D.J.H. (1993) Cosmopolitan biomonitors of trace metals. Marine Pollution Bulletin, 26, 593–601.

[5] Rainbow P.S. (1995) Biomonitoring of heavy metal availability in the marine environment. Marine Pollution Bulletin, 31, 183–192.

[6] Conti M.E., Tudino M.B., Muse J.O., Cecchetti G.F. (2002) Biomonitoring of heavy metals and their species in the marine environment: the contribution of atomic absorption spectroscopy and inductively coupled plasma spectroscopy. Research Trends in Applied Spectroscopy, 4, 295–324, Reprinted (or higher parts taken) with a kind permission from Research Trends.

[7] Rai L.C., Gaur J.P., Kumar H.D. (1981) Phycology and heavy-metal pollution. Biological Reviews, 56, 99–151.

[8] Volterra L., Conti M.E., (2000) Algae as biomarkers, bioaccumulators and toxin producers, in: Conti M.E., Botrè F. (Eds.), The Control of Marine Pollution: Current Status and Future Trends. International Journal of Environment and Pollution, 13(1–6), 92–125.

[9] Haritonidis S., Malea P. (1995) Seasonal and local variation of Cr, Ni and Co concentrations in *Ulva rigida* C. Agardh and *Enteromorpha linza* (Linnaeus) from Thermaikos Gulf, Greece. Environmental Pollution, 89, 319–327.

[10] Capone W., Mascia C., Porcu M., Tagliasacchi Masala M.L. (1983) Uptake of lead and chromium by primary producers and consumers in a polluted lagoon. Marine Pollution Bulletin, 14(3), 97–102.

[11] Castagna A., Sinatra F., Castagna G., Stoli A., Zafarana S. (1985) Trace elements evaluations in marine organisms. Marine Pollution Bulletin, 16(10), 416–419.

[12] UNEP/FAO/WHO (1987) Assessment of the state of pollution of the Mediterranean Sea by mercury and mercury compounds. MAP Technical Reports Series No. 18. UNEP, Athens, p. 354.

[13] Muse J.O., Stripeikis J.D., Fernández F.M., d'Huicque L., Tudino M.B., Carducci C.N., Troccoli O.E. (1999) Seaweeds in the assessment of heavy metal pollution in the Gulf San Jorge, Argentina. Environmental Pollution, 104, 315–322.

[14] Munda I.M. (1979) Addition to the check-list of benthic marine algae from Iceland. Botanica Marina, 22, 459–463.

[15] Munda I.M. (1984) Salinity dependent accumulation of Zn, Co and Mn in *Scytosiphon lomentaria* (Lyngb.) Link and *Enteromorpha intestinalis* (L.) Link from the Adriatic Sea. Botanica Marina, 27, 371–376.

[16] Munda I.M., Hudnik V. (1991) Trace metal content in some seaweeds from the northern Adriatic. Botanica Marina, 34, 241–249.

[17] Sawidis T., Voulgaropoulos A.N. (1986) Seasonal bioaccumulation of iron, cobalt and copper in marine algae from Thermaikos Gulf of the northern Aegean Sea, Greece. Marine Environmental Research, 19, 39–47.

[18] Djingova R., Kuleff I., Arpadjan S., Alexerov S., Voulgaropoulos A., Sadwis T. (1987) Neutron activation and atomic absorption of Ulva lactuca and Gracilaria verrucosa from Thermaikos Gulf, Greece. Toxicological and Environmental Chemistry, 15, 149–158.

[19] Ho Y.B. (1990) Metals in *Ulva lactuca* in Hong Kong intertidal waters. Bulletin of Marine Science, 47, 79–85.

[20] Taramelli E., Costantini S., Giordano R., Olivieri N., Perdicaro R. (1991) Cadmium in water, sediments and benthic organisms from a stretch of coast facing the thermoelectric power plant at Torvaldiga (Civitavecchia, Rome), in: UNEP/FAO. Final reports on research projects dealing with bioaccumulation and toxicity of chemical pollutants. MAP Technical Reports Series No. 52, Athens, pp. 15–31.

[21] Catsiki V.A., Papathanassiou E. (1993) The use of chlorophyte *Ulva lactuca* (L.) as indicator organism of metal pollution, in: Proc. Cost. Symp. of Sub Group III. Macroalgae, Eutrophication and Trace Metal Cycling in Estuaries and Lagoons, Thessaloniki, Greece, 93–105.

[22] Schuhmacher M., Domingo J.L., Llobet J.M., Corbella J. (1995) Variations of heavy metals in water, sediments, and biota from the delta of Ebro River, Spain. Journal Environmental Science Health, part a, 30, 1361–1372.

[23] Favero N., Cattalini F., Bertaggia D., Albergoni V. (1996) Metal accumulation in a biological indicator (*Ulva rigida*) from the lagoon of Venice (Italy). Archives Environmental Contamination Toxicology, 31, 9–18.

[24] UNEP (1996) State of the marine and coastal environment in the Mediterranean region. MAP Technical Reports Series No. 100. UNEP, Athens, p. 142.

[25] UNEP/FAO/WHO (1996) Assessment of the state of pollution of the Mediterranean Sea by zinc, copper and their compounds. MAP Technical Reports Series No. 105. UNEP, Athens, p. 288.

[26] Keeney W.L., Breck W.G., Van Loon G.W., Page J.A. (1976) The determination of trace metals in *Cladophora glomerata* as a potential biological monitor. Water Research, 10, 981–984.

[27] Costantini S., Giordano R., Ciaralli L., Beccaloni E. (1991) Mercury, cadmium and lead evaluation in *Posidonia oceanica* and *Codium tomentosum*. Marine Pollution Bulletin, 22, 362–363.

[28] Munda I.M. (1990) Trace metals in Adriatic seaweeds from the Istrian coast, Rapp. P.-V. Rèun. CIESM, 32(1), 4.

[29] Muse J.O., Tudino M.B., d'Huicque L., Troccoli O.E., Carducci C.N. (1989) Atomic absorption spectrometric determination of inorganic and organic arsenic in some marine benthic algae of the Southern Atlantic coasts. Environmental Pollution, 58, 303–312.

[30] Muse J.O., Tudino M.B., d'Huicque L., Troccoli O.E., Carducci C.N. (1995) A survey of some trace elements in seaweeds from Patagonia, Argentina. Environmental Pollution, 87, 249–253.

[31] Nassar C.A.G., Salgado L.T., Yoneshigue-Valentin Y., Amado Filho G.M. (2003) The effect of iron-ore particles on the metal content of the brown alga *Padina gymnospora* (Espírito Santo Bay, Brazil). Environmental Pollution, 123, 301–305.

[32] Bei F., Papathanassiou E., Catsiki V.A. (1990) Heavy metals concentrations in selected marine species from Milos island (Aegean sea), Rapp. International Commission for the Scientific Exploration of the Mediterranean Sea (CIESM), 32, 126.

[33] Conti M.E., Cubadda F., Campanella L. (1998) Metalli in tracce in componenti di un ecosistema marino incontaminato del Mediterraneo: Indagine preliminare. European Countries Biologists Association, Atti del X Congresso Internazionale: Problemi ambientali e sanitari nell'area Mediterranea, Maratea,10–13 Ottobre 1997, 421–429.

[34] Fuge R., James K.H. (1974) Trace metals concentrations in *Fucus* from the Bristol Channel. Marine Pollution Bulletin, 5, 9–12.

[35] Bryan G.W. (1969) The absorption of zinc and other metals by the brown seaweed *Laminaria digitata*. Journal Marine Biological Association UK, 49, 225–243.

[36] Preston A., Jeffries D.F., Dutton J.W.R., Harvey B.R., Steele A.K. (1972) British isles coastal waters: the concentrations of selected heavy metals in sea water, suspended matter and biological indicators – a pilot survey. Environmental Pollution, 3, 69–82.

[37] Barnett B.E., Ashcroft C.R. (1985) Heavy metals in *Fucus vesiculosus* in the Humber estuary. Environmental Pollution, 9, 193–213.

[38] Melhuus A., Seip K.L., Seip H.M., Myklestad S. (1978) A preliminary study of the use of benthic algae as biological indicators of heavy metals pollution in Sorfjorden, Norway. Environmental Pollution, 15, 101–107.

[39] Woolston M.E., Breck W.G., Van Loon G.W. (1982) A sampling study of the brown seaweeds *Ascophillum nodosum* as a marine monitor of trace metals. Water Research, 16, 687–691.

[40] Lecaros O., Soledad Astorga E.M. (1992) Heavy metals in *Macrocystis pyrifera* (giant kelp) from the strait of Magellan coasts. Revista de Biologia Marina, 27(1), 5–16.

[41] Phillips D.J.H. (1990) Use of macroalgae and invertebrates as monitors of metal levels in estuaries and coastal waters, in: Furness R.W., Rainbow P.S. (Eds.), Heavy metals in the Marine Environment. CRC Press, Boca Raton, pp. 81–99.

[42] Langston W.J., Spence S.K. (1995) Biological factors involved in metal concentrations observed in aquatic organisms, in: Tessier A., Turner D.R. (Eds.), Metal Speciation and Bioavailability in Aquatic Systems, IUPAC. John Wiley and Sons Ltd., New York, pp. 407–478.

[43] Harrison R.M. (1987) Physico-chemical speciation and chemical transformations of toxic metals in the environment, in: Coughtrey P.J., Martin M.H., Unsworth M.H. (Eds.), Pollutant transport and fate in ecosystems, Special publication number 5 of the British Ecological Society. Blackwell Scientific Publications, Oxford, pp. 239–247.

[44] Campbell N.A., Reece J.B. (2007) Biology (5th ed.), The Benjamin/Cummings Publishing Company, Inc., San Francisco, 1998.

[45] Jacobs W.P. (1995) *Caulerpa*, Le Scienze, 38, 76–81.

[46] Gutknecht J. (1965) Uptake and retention of cesium 137 and zinc 65 by seaweeds. Limnology and Oceanography, 10, 58–66.

[47] Bryan G.W., Langston W.J., Hummerstone L.G., Burt G.R. (1985) A guide to the assessment of heavy-metal contamination in estuaries using biological indicators. Journal Marine Biological Association UK, Occasional Publications, 4, 1–92.

[48] Seeliger U., Edwards P. (1977) Correlation coefficients and concentration factors of copper and lead in seawater and benthic algae. Marine Pollution Bulletin, 8(1), 16–19.

[49] Luoma S.N. (1983) Bioavailability of trace metals to acquatic organisms – a review. The Science of the Total Environment, 28, 1–22.

[50] Garnham G.W., Codd G.A., Gadd G.M. (1992) Kinetics of uptake and intracellular location of cobalt, manganese and zinc in the estuarine green alga Chlorella salina. Applied Microbiology Biotechnology, 37, 270–276.

[51] Skipnes O., Roald T., Haug A. (1975) Uptake of zinc and strontium by brown algae. Physiologia Plantarum, 34, 314–320.

[52] Depledge M.H., Rainbow P.S. (1990) Models of regulation and accumulation of trace metals in marine invertebrates. Comparative Biochemistry and Physiology, 97C, 1–7.

[53] Ehrlich H.L. (1986) What types of microorganisms are effective in bioleaching, bioaccumulation of metals, ore beneficiation, and desulfurization of fossil fuels? Biotechnology and Bioengineering Symposium, 16, 227–237.

[54] Crist R.H., Oberholser K., Shank N., Nguyen M. (1981) Nature of bonding between metallic ions and algal cell walls. Environmental Science Technology, 15(10), 1212–1217.

[55] Hall A., Fielding A.H., Butler M. (1979) Mechanisms of copper tolerance in the marine fouling alga *Ectocarpus siliculosus*, evidence for an exclusion mechanism. Marine Biology, 54(3), 195–199.

[56] Sorentino C. (1985) Copper resistance in *Hormidium fluitans* (Gay) Heering (Ulotrichaceae. Chlorophyceae). Phycologia, 24(3), 366–368.

[57] Percival E. (1979) The polysaccharides of green, red and brown seaweeds: their basic structure, biosynthesis and function. British Phycological Journal, 14, 103–117.

[58] Overnell J. (1976) Inhibition of marine algal photosynthesis by heavy metals. Marine Biology, 38, 335–342.

[59] Lobban C.S., Harrison P.J., Duncan M.J. (Eds.) (1985) The Physiological Ecology of Seaweeds. Cambridge University Press, New York, p. 242.

[60] Caberi H., Scoullos M. (1996) The role of the green algae *Ulva* in the cycling of copper in marine coastal ecosystems, in: UNEP, Final reports on research projects dealing with eutrophication and heavy metal accumulation. MAP Technical Reports Series No. 104, 83–96.

[61] Gledhill M., Nimmo M., Hill S.J., Brown M.T. (1997) The toxicity of copper (II) species to marine algae, with particular reference to macroalgae. Journal Phycology, 33, 2–11.

[62] Clijsters H., Van Assche F. (1985) Inhibition of photosynthesis by heavy metals. Photosynthesis Research, 7(1), 31–40.

[63] Vymazal J. (1987) Toxicity and accumulation of cadmium with respect to algae and cyanobacteria: a review. Toxicity Assessment, 2(4), 387–415.

[64] Fernández Leborans G., Novillo A. (1996) Toxicity and bioaccumulation of cadmium in *Olisthodiscus Luteus* (Raphidophyceae). Water Research, 30(1), 57–62.

[65] Devi Prasad P.V., Devi Prasad P.S. (1982) Effect of cadmium, lead and nickel on three freshwater green algae. Water, Air, Soil Pollution, 17(3), 263–268.

[66] Çalgan D.T., Yenigün O. (1996) Toxicity of arsenic, cadmium and nickel on the cyanobacterium *Anabaena Cylindrica*. Environmental Technology, 17(5), 533–540.

[67] Gaur J.P., Rai L.C. (2001) Heavy metal tolerance in algae, in: Rai L.C., Gaur J.P. (Eds.), Algal Adaptation to Environmental Stresses: Physiological, Biochemical and Molecular Mechanisms. Berlin, Springer, pp. 363–388.

[68] Zhou W., Qiu B. (2004) Mechanisms for heavy metal detoxification and tolerance in algae. Hupo Kexue, 16(3), 265–272.

[69] Takamura N., Kasai F., Watanabe M.M. (1990) Unique response of Cyanophyceae to copper. Journal Applied Phycology, 2, 293–296.

[70] Hirata K., Tsujimoto Y., Namba T., Ohta T., Hirayanagi N., Miyasaka H., Zenk M.H., Miyamoto K. (2001) Strong induction of phytochelatin synthesis by zinc in marine green alga, *Dunaliella tertiolecta*. Journal of Bioscience and Bioengineering, 92(1), 24–29.

[71] Ahner B.A., Morel F.M.M. (1995) Phytochelatin production in marine algae. 2. Induction by various metals. Limnology Oceanography, 40(4), 658–665.

[72] Söderlund S., Forsberg A., Perdersen M. (1988) Concentrations of cadmium and other metals in Fucus vesiculosus L. and Fontinalis delecarlica Br. Eur. from the Northern Baltic Sea and the Southern Bothnian sea. Environmental Pollution, 51, 197–212.

[73] Barreiro R., Picado L., Real C. (2002) Biomonitoring heavy metals in estuaries: a field comparison of two brown algae species inhabiting upper estuarine reaches. Environmental Monitoring and Assessment, 75(2), 121–134.

[74] Burridge T.R., Bidwell J. (2002) Review of the potential use of brown algal ecotoxicological assays in monitoring effluent discharge and pollution in southern Australia. Marine Pollution Bulletin, 45(1–12), 140–147.

[75] Mecozzi M., Conti M.E. (2007) A new approach based on polynomial functions for measuring the acute toxicity of environmental samples. International Journal of Environment and Health, 1(1), 87–97.

[76] Mallick N. (2003) Biotechnological potential of *Chlorella vulgaris* for accumulation of Cu and Ni from single and binary metal solutions. World Journal of Microbiology & Biotechnology, 19(7), 695–701.

[77] Awasthi M., Rai L.C. (2004) Adsorption of nickel, zinc and cadmium by immobilized green algae and cyanobacteria: a comparative study. Annals of Microbiology (Milano, Italy), 54(3), 257–267.

[78] Ruangsomboon S., Wongrat L. (2006) Bioaccumulation of cadmium in an experimental aquatic food chain involving phytoplankton (*Chlorella vulgaris*), zooplankton (*Moina macrocopa*), and the predatory catfish *Clarias macrocephalus x C. gariepinus*. Aquatic Toxicology, 78(1), 15–20.

[79] Bajguz A., Godlewska-Zylkiewicz B. (2004) Protective role of 20-hydroxyecdysone against lead stress in *Chlorella vulgaris* cultures. Phytochemistry, 65(6), 711–720.

[80] Hao S., Xiaorong W., Qin W., Liansheng W., Yijun C., Iemei D., Zhong L., Mi C. (1998) The species of spiked rare earth elements in sediment and potential bioavailability to algae (*Chlorella Vulgarize Beijerinck*). Chemosphere, 36(2), 329–337.

[81] Darnall D.W., Greene B., Henzl M.T., Hosea J.M., Mcpherson R.A., Sneddon J., Alexeer M.D. (1986) Selective recovery of gold and other metal ions from an algal biomass. Environmental Science Technology, 20, 206–208.

[82] Rehman A., Shakoori A.R. (2001) Heavy metal resistance *Chlorella spp.*, isolated from tannery effluents, and their role in remediation of hexavalent chromium in industrial waste water. Bulletin of Environmental Contamination and Toxicology, 66(4), 542–547.

[83] Yoshida N., Ishii K., Okuno T., Tanaka K. (2006) Purification and characterization of cadmium-binding protein from uniceluar alga *Chlorella sorokiniana*. Current Microbiology, 52(6), 460–463.

[84] Conti M.E. (1998) Indagine sull'accumulo di metalli pesanti in un indicatore biologico (*Ulva lactuca*): risultati preliminari. Rassegna Chimica, 4, 390–394.

[85] Haritonidis S., Malea P. (1999) Bioaccumulation of metals by the green alga *Ulva rigida* C. Agardh from Thermaikos Gulf, Greece. Environmental Pollution, 104, 365–372.

[86] Phaneuf D., Cote I., Dumas P., Ferron L.A., LeBlanc A. (1999) Evaluation of the contamination of marine algae (Seaweed) from the St. Lawrence River and likely to be consumed by humans. Environmental Research, 80(2 Pt 2), S175–S182.

[87] Sawidis T., Brown M.T., Zachariadis G., Sratis I. (2001) Trace metal concentrations in marine macroalgae from different biotopes in the Aegean sea. Environmental International, 27, 43–47.

[88] Villares R., Puente X., Carballeira A. (2001) Ulva and Enteromorpha as indicators of heavy metal pollution. Hydrobiologia, 462, 221–232.

[89] Storelli M.M., Storelli A., Marcotrigiano G.O. (2001) Heavy metals in the aquatic environment of the Southern Adriatic Sea, Italy. Macroalgae, sediments and benthic species. Environment International, 26, 505–509.

[90] Caliceti M., Argese E., Sfriso A., Pavoni B. (2002) Heavy metal contamination in the seaweeds of the Venice lagoon. Chemosphere, 47, 443–454.

[91] El-Moselhy K.M., Gabal M.N. (2004) Trace metals in water, sediments and marine organisms from the northern part of the Gulf of Suez, Red Sea. Journal of Marine Systems, 46, 39–46.

[92] Kaimoussi A., Mouzdahir A., Saih A. (2004) Variations saisonnières des teneurs en métaux (Cd, Cu, Fe, Mn et Zn) chez l'algue *Ulva lactuca* prélevée au niveau du littoral de la ville d'El Jadida (Maroc). Comptes Rendus Biologies, 327, 361–369.

[93] Eide I., Myklestad S., Melsom S. (1980) Long-term uptake and release of heavy metals by *Ascophyllum Nodosum* (L.) Le Jol. (Phaeophyceae) in situ. Environmental Pollution (Series A), 23, 19–28.

[94] Myklestad S., Eide I., Melsom S. (1978) Exchange of heavy metals in *Ascophyllum nodosum* (L.) Le Jol. in situ by means of transplanting experiments. Environmental Pollution, 16(4), 277–284.

[95] Sawidis T.H., Voulgaropoulos A.N. (1986) Seasonal bioaccumulation of iron, cobalt and copper in marine algae from Thermaikos Gulf of the northern Aegean sea, Greece. Marine Environmental Research, 19, 39–47.

[96] Viaroli P., Bartoli M., Azzoni R., Giordani G., Mucchino C., Naldi M., Nizzoli D., Taje L. (2005) Nutrient and iron limitation to *Ulva* blooms in a eutrophic coastal lagoon (Sacca di Goro, Italy). Hydrobiologia, 550, 57–71.

[97] Campanella L., Conti M.E., Cubadda F., Sucapane C. (2001) Trace metals in seagrass, algae and molluscs from an uncontaminated area in the Mediterranean. Environmental Pollution, 111, 117–126.

[98] Philips D.J.H. (1994) Macrophytes as biomonitors of trace metals, in: Kramer K.J.M. (Ed.), Biomonitoring of Coastal Waters and Estuaries. CRC Press, Boca Raton, pp. 85–103.

[99] Maserti B.E., Ferrara R., Paterno P. (1988) Posidonia as an indicator of mercury contamination. Marine Pollution Bulletin, 19(8), 381–382.

[100] Malea P., Haritonidis S. (1989) Uptake of copper, cadmium, zinc and lead in *Posidonia oceanica* (Linnaeus) from Antikyra Gulf, Greece: preliminary note. Marine Environmental Research, 28(1–4), 495–498.

[101] Sanchiz C., Benedito V., Pastor A., Garcia-Carrascosa A.M. (1990) Bioaccumulation of heavy metals in *Posidonia oceanica* (L.) Delile and *Cymodocea nodosa* (Ucria) Aschers. at an uncontaminated site in the east coast of Spain, Rapp. International Commission for the Scientific Exploration of the Mediterranean Sea (CIESM), 32(1), 13.

[102] Warnau M., Ledent G., Temara A., Bouquegneau J.M., Jangoux M., Dubois P. (1995) Heavy metals in *Posidonia oceanica* and *Paracentrotus lividus* from seagrass beds of the north-western Mediterranean. Science of the Total Environment, 171(1–3), 95–99.

[103] Warnau M., Fowler S.W., Teyssie J.L. (1996) Biokinetics of selected heavy metals and radionuclides in two marine macrophytes: the seagrass *Posidonia oceanica* and the alga *Caulerpa taxifolia*. Marine Environmental Research, 41(4), 343–362.

[104] Conti M.E., Facchini M., Botrè F. (2000) Indicadores Biológicos para la evaluación de metales pesados en un ecosistema marino del mediterraneo. Acta Toxicológica Argentina, 8(2), 65–68.

[105] Conti M.E., Sucapane C. (1998) La qualità degli ecosistemi marini: bio-monitoraggio di metalli pesanti in un sito incontaminato. Atti del XVIII Congresso Nazionale di Merceologia, Verona 1–3 ottobre, 143–150.

[106] Gosselin M., Bouquegneau J.M., Lefebvre F., Lepoint G., Pergent G., Pergent-Martini C., Gobert S. (2006) Trace metal concentrations in *Posidonia oceanica* of North Corsica (northwestern Mediterranean Sea): use as a biological monitor? BMC Ecology, 6, art. 12.

[107] UNEP (1999) Proceedings of the Workshop on Invasive *Caulerpa* species in the Mediterranean, Heraklion, Crete, Greece, 18–20 March 1998. MTS N° 125, UNEP, Athens 1999, p. 317.

[108] Meinesz A., Hesse B. (1991) Introduction et invasion de l'algue tropicale *Caulerpa taxifolia* en Méditerranée Occidentale. Oceanologica Acta, 14, 415–426.

[109] International *Caulerpa taxifolia* Conference Proceedings (2002) January 31–February 1, San Diego, California U.S.A., University of California, Williams E., Grosholz E. (Eds.), p. 245.

[110] Catsiki V.A., Panayotidis P. (1993) Copper, chromium and nickel in tissues of the Mediterranean seagrasses *Posidonia oceanica* and *Cymodocea nodosa* (Potamogetonaceae) from Greek coastal areas. Chemosphere, 26(5), 963–978.

[111] Schlacher-Hoelinger M.A., Schlacher T.A. (1998) Differential accumulation patterns of heavy metals among the dominant macrophytes of a Mediterranean seagrass meadow. Chemosphere, 37, 1511–1519.

[112] Pergent-Martini C., Pergent G. (2000) Marine phanerogams as a tool in the evaluation of marine trace-metal contamination: an example for the Mediterranean, in: Conti M.E., Botrè F. (Eds.), The control of marine pollution. Current status and future trends. International Journal of Environment and Pollution, 13(1–6), 126–147.

[113] Brix H., Lyngby J.E. (1984) A survey of the metallic composition of *Zostera marina* (L.) in the Limfjord, Denmark. Archives of Hydrobiology, 99, 347–359.

[114] Brix H., Lyngby J.E. (1989) Heavy metals in eelgrass (*Zostera marina* L.) during growth and decomposition. Hydrobiology, 176/177, 189–196.

[115] Ward T.J. (1987) Temporal variation of metals in the seagrass *Posidonia australis* and its potential as a sentinel accumulator near a lead smelter. Marine Biology, 95, 315–321.

[116] Brix H., Lyngby J.E., Schierup H.-H. (1983) The reproducibility in the determination of heavy metals in marine plant material–an interlaboratory calibration. Marine Chemistry, 12, 69–85.

[117] Ledent G., Mateo M.A., Warnau M., Temara A., Romero J., Dubois P. (1995) Element losses following distilled water rinsing in leaves of the seagrass *Posidonia oceanica* (L.) Delile. Aquatic Botany, 52, 229–235.

[118] Bond A.M., Bradbury J.R., Hudson H.A., Garnham J.S., Hanna P.J., Stoher S. (1985) Kinetic studies of lead (II) uptake by the seagrass *Zostera muelleri* in water by radiotracing, atomic absorption spectrometry and electrochemical techniques. Marine Chemistry, 16, 1–9.

[119] Conti M.E., Botrè F. (2001) Honey bees and their products as potential bioindicators of heavy metals contamination. Environmental Monitoring & Assessment, 69, 267–282.

[120] Conti M.E., Ciprotti L. (1999) Il miele come bioindicatore della contaminazione da metalli pesanti. Rivista Sanità Rapporti, 4(1), 75–84.

[121] Conti M.E., Saccares S., Cubadda F., Cavallina R., Tenoglio C.A., Ciprotti L. (1998) Il miele nel Lazio: Indagine sul contenuto in metalli in tracce e radionuclidi. La Rivista di Scienza dell'Alimentazione, 2, 107–119.

[122] Cubadda F., Conti M.E., Campanella L. (2001) Size-dependent concentrations of trace metals in four Mediterranean gastropods. Chemosphere, 45(4), 561–569.

[123] Conti M.E., Stripeikis J., Iacobucci M., Cucina D., Cecchetti G., Tudino M.B. (2006) Trace metals in molluscs from the Beagle Channel (Argentina): a preliminary study, in: Brebbia C.A., Conti M.E., Tiezzi E. (Eds.), Management of Natural Resources, Sustainable Development and Ecological Hazards – The Ravage of the Planet Conference 2006. WITpress, Boston, pp. 473–483.

[124] Brown M.T., Depledge M.H. (1998) Determinants of trace metal concentrations in marine organisms, in: Langston W.J., Bebianno M.J. (Eds.), Metal Metabolism in Acquatic Environments. Chapman & Hall, UK, pp. 185–217.

[125] Marigomez I., Soto M., Cajaraville M.P., Angulo E., Giamberini L. (2002) Cellular and subcellular distribution of metals in molluscs. Microscopy Research and Technique, 56(5), 358–392.

[126] Rauser W.E. (1995) Phytochelatins and related peptides. Structure, biosynthesis, and function. Plant Physiology, 109, 1141–1149.

[127] Ahner B.A., Price N.M., Morel F.M.M. (1994) Phytochelatin production by marine phytoplankton at low free metal ion concentrations: laboratory studies and field data from Massachusetts Bay. Proceedings of the National Academy of Sciences of the USA, 91, 8433–8436.

[128] Langston W.J., Bebianno M.J., Burt G.R. (1998) Metal handling in molluscs, in: Langston W.J., Bebianno M.J. (Eds.), Metal Metabolism in Acquatic Environments. Chapman & Hall, UK, pp. 219–283.

[129] Boyden C.R. (1974) Trace element content and body size in molluscs. Nature, 251(5473), 311–314.

[130] Boyden C.R. (1977) Effect of size upon metal content of shellfish. Journal of the Marine Biological Association of the United Kingdom, 57(3), 675–714.

[131] Conti M.E., Cucina D., Mecozzi M. (2007) Regression analysis model applied to biomonitoring studies. Environmental Modeling Assessment, DOI: 10.1007/s10666-007-9113-7 (*in press*).

[132] Saavedra Y., Gonzalez A., Fernandez P., Blanco J. (2004) The effect of size on trace metal levels in raft cultivated mussels (*Mytilus galloprovincialis*). Science of the Total Environment, 318(1–3), 115–124.

[133] Catsiki V.A., Vakalopoulou C., Moraitou-Apostolopoulou M., Verriopoulos G. (1993) *Monodonta turbinata* (Born); toxicity and bioaccumulation of Cu and copper + Chromium mixtures. Toxicological and Environmental Chemistry, 37(3–4), 173–184.

[134] Foster P., Chacko J. (1995) Minor and trace elements in the shell of *Patella vulgata* (L.). Marine Environmental Research, 40(1), 55–76.

[135] Conti M.E., Cecchetti G. (2003) A biomonitoring study: trace metals in algae and molluscs from Tyrrhenian coastal areas. Environmental Research, 93(1), 99–112.

[136] Conti M.E., Iacobucci M., Cecchetti G. (2007) A biomonitoring study: trace metals in seagrass, algae and molluscs in a marine reference ecosystem (southern Tyrrhenian sea). International Journal of Environment and Pollution, 29(1–3), 308–332.

[137] Conti M.E., Iacobucci M., Mangani G., Cecchetti G. (2003) I molluschi quali indicatori biologici per l'area Mediterranea. Acqua & Aria, 9, 96–102.

[138] Conti M.E., Iacobucci M., Cecchetti G. (2004) The biomonitoring approach as a tool of trace metal assessment in an uncontaminated marine ecosystem: the Island of Ustica (Sicily, Italy), in: Martin-Duque J.F., Brebbia C.A., Godfrey A.E., Díaz de Terán J.R. (Eds.), Monitoring, Simulation and Remediation of the Geological Environment. WIT Press, Boston, pp. 335–344.

[139] Conti M.E., Iacobucci M., Cecchetti G. (2005) A statistical approach applied to trace metal data from biomonitoring studies. International Journal of Environment and Pollution, 23(1), 29–41.

[140] Schlacher-Hoenlinger M.A., Schlacher T.A. (1998) Accumulation, contamination, and seasonal variability of trace metals in the coastal zone – patterns in a seagrass meadow from the Mediterranean. Marine Biology, 131, 401–410.

[141] Caredda A.M., Cristini A., Ferrara C., Lobina M.F., Baroli M. (1999) Distribution of heavy metals in the Piscinas beach sediments (SW Sardinia, Italy). Environmental Geology, 38(2), 91–100.

[142] Sanchiz C., Garcia-Carrascosa A.M., Pastor A. (2000) Heavy metal contents in soft-bottom marine macrophytes and sediments along the Mediterranean coast of Spain. Marine Ecology, 21(1), 1–14.

[143] Kljaković-Gašpić Z., Antolić B., Zvonarić T., Barić A. (2004) Distribution of cadmium and lead in *Posidonia oceanica* (L.) Delile from the middle Adriatic Sea. Fresenius Environmental Bulletin, 13(11b, y), 1210–1215.

[144] Tranchina L., Bellia S., Brai M., Hauser S., Rizzo S., Bartolotta A., Basile S. (2004) Chemistry, mineralogy and radioactivity in *Posidonia Oceanica* meadows from north-western Sicily. Chemistry and Ecology, 20, 203–214.

[145] Amado Filho G.M., Andrade L.R., Karez C.S., Farina M., Pfeiffer W.C. (1999) Brown algae species as biomonitors of Zn and Cd at Septiba Bay, Rio de Janeiro, Brazil. Marine Environmental Research, 48, 213–224.

[146] Sánchez-Rodríguez I., Huerta-Díaz M.A., Choumiline E., Holguín-Quiñones O., Zertuche-Gonzàlez J.A. (2001) Elemental concentrations in different species of seaweeds from Loreto Bay, Baja California Sur, Mexico: implications for the geochemical control of metals in algal tissue. Environmental Pollution, 114, 145–160.

[147] Topcuoğlu S., Güven K.C., Balkis N., Kirbaşoğlu Ç. (2003) Heavy metal monitoring of marine algae from the Turkish Coast of the Black Sea, 1998–2000. Chemosphere, 52, 1683–1688.

[148] Turkmen M., Turkmen A., Akyurt I., Tepe Y. (2005) Limpet, *Patella caerulea*, Linnaeus, 1758 and barnacle, *Balanus* sp., as biomonitors of trace metal availabilities in Iskenderun Bay, northern East Mediterranean Sea. Bulletin of Environmental Contamination and Toxicology, 74(2), 301–307.

[149] Nicolaidou A., Nott J.A. (1998) Metals in sediment, seagrass and gastropods near a nickel smelter in Greece: possible interactions. Marine Pollution Bulletin, 36(5), 360–365.

[150] Blackmore G. (2001) Interspecific variation in heavy metal body concentrations in Hong Kong marine invertebrates. Environmental Pollution, 114, 303–311.

5   Lichens as bioindicators of air pollution

M E Conti

1   Introduction

Lichens are considered the result of a symbiotic association of a fungus and an alga. More precisely the term "alga" indicates either a Cyanobacteriae or a Chlorophyceae; the fungus is usually an Ascomycetes, although on rare occasions it may be either a Basidiomycetes or a Phycomycetes.

In this association, the alga is the part that is occupied with the formation of nutrients, since it contains chlorophyll (Chl), while the fungus supplies the alga with water and minerals. These organisms are perennial and maintain a uniform morphology over time. They grow slowly, have a large-scale dependence upon the environment for their nutrition, and – differently from vascular plants – they do not shed parts during growth. Furthermore, their lack of cuticle or stoma means that the different contaminants are absorbed over the entire surface of the organism [1–3].

As far back as 1866, a study was published on epiphytic lichens used as bioindicators [4]. Lichens are the most studied bioindicators of air quality [5]. They have been defined as "permanent control systems" for air pollution assessment [6].

During the last 30 years, many studies have stressed the possibility of using lichens as biomonitors of air quality in view of their sensitivity to various environmental factors, which can provoke changes in some of their components and/or specific parameters [3, 7–30].

For indeed, many physiological parameters are used to evaluate environmental damage to lichens, such as: photosynthesis [31–33]; chlorophyll content and degradation [34–37]; decrease of ATP; variations in respiration levels [34]; changes in the level of endogenous auxins; and ethylene production [38–40].

Furthermore, laboratory exposure to SO_2 causes relevant membrane damage to lichen cells [41]. Many studies show a positive correlation between the sulphur content of lichens and SO_2 present in the atmosphere [42–45].

Various authors report that the concentration of Chl $a + b$ is altered by vehicle traffic pollution [46–48], and by urban emissions [49]. In general, lichens that are transplanted into areas with intense vehicle traffic show an increase in Chl $a + b$ concentration that is proportional to increases in emissions. Such effects are generally caused by traffic emissions and in particular, sulphur and nitrogen oxides. In areas with intense vehicle traffic and elevated levels of industrial pollution, high values are obtained for Chl *b*/Chl *a* ratios.

Air traffic, and in particular the effects of kerosene and benzene, seems to have a lesser effect on the lichen population than vehicle traffic. This has been demonstrated in a study of Hamburg airport [50].

Lichens may be used as bioindicators and/or biomonitors in two different ways [3, 23, 51, 52]:

i. by mapping all species present in a specific area (method A);
ii. through the individual sampling of lichen species and measurement of the pollutants that accumulate in the thallus; or by transplanting lichens from an uncontaminated area to a contaminated one, then measuring the morphological changes in the lichen thallus and/or evaluating the physiological parameters and/or evaluating the bioaccumulation of the pollutants (method B).

2 Lichens in the control of environmental contamination

2.1 The Index of Atmospheric Purity (IAP)-method (method A)

The compositional changes in lichen communities are correlated with changes in levels of atmospheric pollution. The application of method A allows the elaboration of an Index of Atmospheric Purity (IAP). This method [53]makes it possible to map out the quality of the air in a determined area. The IAP gives an evaluation of the level of atmospheric pollution, which is based on the number (*n*), frequency (*F*) and tolerance of the lichens present in the area under study. There are 20 different formulae for IAP calculation, and these are able to predict, to a good level of approximation, the degrees of eight atmospheric pollutants measured using automatic control stations (SO_2, NO_x, Cl, Cd, Pb, Zn and dusts) [54].

The formula with the highest correlation with pollution data is that which considers as a parameter only the frequency (*F*) of the lichen species present in a sampling network comprising 10 areas [3]:

$$IAP = \sum Fi$$

F is the frequency (max. 10) of every *i*th species that is calculated as the number of rectangles in the grid (a rectangle of the dimensions 30 × 50 cm, subdivided into 10 areas measuring 15 × 10 cm each), in which a given species appears [54]. It has been shown that the frequency method makes it possible to predict pollution levels with a certainty of over 97% [55–57].

Method A foresees a choice of sampling stations on the basis of the presence of suitable trees on which it is possible to observe lichens. The difficulty of this method lies in finding the same tree species in the study sites so as to enable homogenous observations to be made. For example, in Italy trees of the *Tilia*, *Acer*, *Quercus* and other species are used. In the event that the species are not totally homogenous, observations can be made using other, different species of tree. When selecting suitable trees, it is necessary to take into account the state of damage to the bark as well as the inclination of the trunk (this must be < 10%) and the circumference (min. 70 cm).

Periodic notations are made of all lichen species present within the network (made on a weekly and monthly basis, etc.). A frequency value (F) is given for each species noted and this corresponds to the number of sub-units within the network in which it is present (minimum = 1, maximum = 10). The IAP is then computed for each tree and each station. Is also relevant to reduce the many possible sources of variability affecting lichen diversity [58].

The values obtained may be plotted in order to create an air quality map. The IAP values are grouped into five quality levels which are given in table 1 [59, 60].

The main part of the studies that concern air quality in Italy deal fundamentally with atmospheric pollution in towns and cities or in larger geographical areas, where different sites with different impacts are compared. Of the numerous works, we quote as examples data collected in different Italian sites: the city of Isernia [61]; the province of Potenza [56]; the cities of Trieste [62], Udine [63], Pistoia [64], Siena [65], Montecatini Terme [66], Trento [57] and La Spezia [67,68]; the Veneto region [69]; the Valle del Susa in Piedmont region [70]; the city of Teramo [71], the volcanic areas of Italy [72]; the city of Pavia [73] and the province of Viterbo [74].

Cislaghi and Nimis [75] report a high degree of correlation between lung cancer and the biodiversity of lichens as a result of atmospheric pollution. These conclusions are based upon thousands of observations made in 662 sites in the Veneto region (northern Italy). These high correlation levels have been found for the more common atmospheric pollutants, such as SO_2, NO_3, dusts and SO_4^{2-}, which is respectively: $r^2 = 0.93$, 0.87, 0.86 and 0.85: $p < 0.01$ in all cases.

Of the many lichen species present [76], *Physcia tenella* is among the most common in Italy, above all below the mountain areas. It is a species that is considered to be toxin-tolerant, even if experts are in disagreement as to its

Table 1: Quality levels of Index of Atmospheric Purity (IAP).

Level A	$0 \leq IAP \leq 12.5$	Very high level of pollution
Level B	$12.5 < IAP \leq 25$	High level of pollution
Level C	$25 < IAP \leq 37.5$	Moderate level of pollution
Level D	$37.5 < IAP \leq 50$	Low level of pollution
Level E	$IAP > 50$	Very low level of pollution

sensitivity to sulphur dioxide. In urban environments or cultivated areas, it is possible to find toxin-tolerant species that belong – from a phyto-sociological viewpoint – to the lichen association category of *Xanthorion parietinae* [77]. The more toxin-tolerant species include *Pheophyscia orbicularis* and *Candelaria concolor*; this latter being present at the limit of the "lichen desert".

Different lichen species react to different pollutants in different ways, and various authors give lists that classify them according to sensitivity [76]. The classification of lichen species is one of the most discussed points in literature. In particular, sensitivity to SO_2 is the base factor for most classifications. Several authors, however, suggest classification on the basis of a scale of semiquantitative characteristics [78]; while others classify lichen species according to SO_2 sensitivity on a scale that distinguishes between "acid" and "eutrophic" bark [79].

The latter method, basically qualitative, considers the degree of atmospheric pollution varying from 10 (zero pollution pure air) to 0 (strongest pollution) as a function of SO_2 levels. Each level is defined by various epiphytic lichens of broad ecological amplitude grouped in different communities according to the acidic or alkaline character of the bark. This method, due to its rapidity and sensitivity, can be applied to draw cartographic representations of pollution indexes on ample geographic areas, also based on absolute values, provided the lichen flora is comparable to that used for the original reference study in United Kingdom. A partial drawback of this method is that the knowledge of 80 lichens species is required [80].

Van Haluwyn and Lerond [81] proposed a qualitative method based on lichenosociology. The authors suggest the use of a 7-level scale (indicated by letters A to G) defined on the basis of easily recognisable species. According to this method, letters A to E refer to strongly polluted, and letters F–G to less polluted areas. Studies performed on Northern France showed that the two groups of zones relates to SO_2 levels higher and lower than 30 $\mu g\ m^{-3}$, respectively. According to this method, even the presence of one species only can be sufficient to characterise one zone. One of the major advantages of this method is that it is not directly correlated to the levels of SO_2 only, but it is based on the overall response capacity of epiphytic communities as a whole [3, 82].

The IAP method is the most commonly adopted in some European countries. Despite the quantitative information that it can supply, this method also presents some disadvantages, primarily among them the fact that a deep knowledge of lichen flora is required, and that it refers to a specific group of environmental pollutants [54].

More recently, a Lichen Biodiversity Index (IBL) [83] has been proposed in Italy. This method tends to improve some characteristics of the sampling procedures such as the choice of sampling sites, the trees and the position of sampling grids.

Another quantitative method is the index of poleotolerance (IP) proposed by Trass [84] and subsequently reviewed by Deruelle [80]. This method, developed and applied in Estonia, allows to draw a map of pollution on the basis of a mathematical index that, in turns, is obtained following observation carried out in predefined conditions. This method considers trees of different age and species.

Every observation is performed taking into account the area of the bark covered by lichens, which is related to a graded reference scale varying from 1 to 10 according to the per cent of covered surface. Indeed, every species is classified according to the IP, calculated as [3, 39]:

$$IP = \sum_{1}^{n} a_i \frac{c_i}{C_i}$$

where n represents the number of considered species, a_i the degree of tolerance of each species, c_i the corresponding level of covering and C_i the overall degree of covering of all species as a whole.

According to this method, an IP value of 10 refers to a zone of lichen desert, while normal condition correspond to levels 1–4. IP can also be correlated with SO_2 levels, where an IP of 1–2 corresponds to zero SO_2 and an IP of 10 to SO_2 concentration higher than 300 $\mu g\ m^{-3}$.

Deruelle; Lerond et al.; and Van Haluwyn and Lerond [80, 82, 85] critically reviewed the main advantages and disadvantages of the above-mentioned qualitative and quantitative methods. The two qualitative methods are also compared by Khalil and Asta [86], in a French study considering the recolonisation by pollution sensitive lichens of the Lyon area.

Some authors [87] propose a methodology for biomonitoring climatic changes by measuring the lichen communities of calcareous rocks and for determining the Trend Detection Index (TDI) with which to verify the sum of lichen species, allowing variability coefficients to be applied to lichen communities that are sensitive to average annual temperature changes of up to 0.8 °C. This application is of great interest, above all taking into account that realistic predictions for planetary global warming should be in the range of 2.5 °C for the next decades.

2.2   Use of native lichens and the transplant method (method B)

In areas where lichens are not killed by contaminants, it is possible to make biomonitoring studies through the direct analysis of contaminants in the thallus. Method B, which consists of transplanting lichen thalli, has the great advantage of being applicable even in "lichen desert" areas (in areas that are unsuited to lichen survival due to high pollution levels), or it can be used in areas where there are no suitable substrata [88, 89].

The lichen thalli used are taken from tree bark in areas of low pollution and then fixed to suitable surfaces (e.g. cork) and exposed in monitoring areas where samples are taken periodically in order to evaluate the health of the thalli and their degree of damage. Lichen damage is expressed as a percentage of necrotised lichen surface.

The main problem with this method is that found in the difficulty in providing a valid interpretation of transplanted thalli damage percentages. There are also methods that allow identification of necrotic areas, defining them on photographs of lichen thalli. A certain error margin, due to the subjective interpretation of the images, has also been found for this procedure. Possible tendencies to over- or underestimate may be corrected through use of statistics tests (χ^2, t student).

The transplant method is also used in classical bioaccumulation studies that analyse contaminants in tissue. Numerous works regarding this method are concerned with trace elements and in particular, bioaccumulation, absorption, retention, localisation and release, tolerance and toxicity [3, 88, 90–118].

2.3 Heavy metals and trace elements

The accumulation of metals in plants depends upon many factors, such as the availability of elements; the characteristics of the plants, such as species, age, state of health, type of reproduction, etc.; and other such parameters as temperature, available moisture, substratum characteristics, etc. [3, 119]. Contaminants deposit on lichens through normal and indirect (occult) precipitation. This latter includes mist, dew, dry sedimentation and gaseous absorption [120]. Indirect precipitation occurs in highly stable atmospheric conditions and contains higher nutriment and contaminant concentrations of different orders of size when compared to normal precipitation [121].

In general, three mechanisms have been put forward with regard to the absorption of metals in lichens [103]:

1. intracellular absorption through an exchange process;
2. intracellular accumulation;
3. entrapment of particles that contain metals.

Many experts have attempted to increase knowledge of these bonding processes – that is, the interaction between lichen and metal – using various analytical techniques, such as nuclear magnetic resonance (NMR), electron paramagnetic resonance (EPR) and luminescence. It should, however, be noted that knowledge regarding the understanding of the entire process that is responsible for metal absorption and accumulation in lichens is still scarce. A new approach has been attempted [122], where metal–lichen interaction is studied by applying microcalorimetric techniques with the aim of obtaining enthalpic measurement data. To carry out these tests and to process the microcalorimetric data, the metal–lichen complex is considered as an overall co-ordinating agent, given that it is not possible at this time to know which particular molecule is responsible for co-ordination with the metal. Considering the constant towards equilibrium and the enthalpy trend for *Evernia prunastri*, the following trend has been found: Pb $\gg$ Zn $>$ Cd $\approx$ Cu $\approx$ Cr; which indicates a good correlation between the metal bond and the enthalpy values in the absorption process (metal uptake).

Lichens are also excellent bioaccumulators of elements and trace elements, since the concentrations found in their thalli can be directly correlated with those in the environment [123–133].

Studies made of transplanted *Evernia prunastri* highlight the fact that the capacity for Pb accumulation expressed as the relationship between the concentration in the latest sample and the initial concentration value, is 10.2 in the Fontainbleau site (France), 3.7 for the Würzburg site (Germany) [101] and 4.4 for the city of Rome (Italy) [134]. Recently, Conti *et al.* [88] also employed *Evernia prunastri* for biomonitoring the atmospheric deposition of heavy metals at urban, rural and industrial sites in Central Italy. Lichen samples were collected in a control site 1500 m a. s. l. (Parco Nazionale d'Abruzzo, Central Italy) and subsequently transplanted at urban site (Cassino city centre), at rural location (7 km away from Cassino city) and at industrial location (Piedimonte S. Germano) surrounding an automobile factory. Lichen samples were transplanted at the four cardinal points of each site. Studies of bioaccumulation of Pb, Cd, Cr, Cu and Zn in lichen samples were performed five times at regular intervals between November 2000–December 2001. Results showed the good ability of *Evernia prunastri* to accumulate the heavy metals under study. As expected, the area chosen as control site showed significantly (Friedman test, cluster analysis) lower impact in comparison to the other sites and the rural site showed smaller impact than the urban and the industrial sites.

More recently, a study dealing with nuclear microprobe analysis of an *E. prunastri* transplanted thallus in thin cross-sections [118], concluded that trace elements are mainly concentrated on the cortex of the thallus, with the exception of Zn, Ca and K.

In Italy, different biomonitoring studies carried out using lichens have shown that Pb is still very widespread in spite of the introduction of lead-free petrol. This indicates that high levels of this metal are still released (and/or re-suspended) by vehicle traffic [65, 135, 136]. Vehicle traffic seems to be the main source of atmospheric Cr, Cu and Pb in the central Italian sites [71].

Climatic factors most probably play an important role in the bioaccumulation of heavy metals, even if this role is as yet unclear. The direction in which pollutants are transported by the wind is most surely fundamental in determining their main fallout points. Nimis *et al.* [6] correlates pollution from an industrial pole (northern Italy) with that at a distant agricultural centre, situated in the predominant wind direction.

It is well known that heavy metal content in lichen thalli tends to alternate over time in phases of accumulation and subsequent release. The causes of these differences may lie in the incidence on this phenomenon of acid rain. Deruelle [101] indicates that the periodic releases of Pb that occur in lichens may depend upon lixiviation induced by acid precipitation. Indeed, laboratory experiments show that lixiviation does not occur at pH 7 [6]. Heavy metals do, in any case, influence water loss in lichen thalli, and the accumulative effect of Pb, Cu and Zn on water loss, after absorption of a mix of metals in solution, has been observed in the laboratory [137].

Moreover, laboratory studies showed that the number of mobile hydrogen ions bonded in lichens (*Hypogymnia physodes*) depends on the concentration of hydrogen ions in the precipitation with which the lichens are in contact and the type and concentration of other cations contained in the precipitation [138]. The number of hydrogen ions accumulated in lichens is proportional to the level of atmospheric precipitation acidification and the pH of precipitation should be determined by the assessment of mobile cations bonded in exposed or naturally grown lichens [138, 139].

Altitude seems to play an important role in Pb and Cd concentrations, as studied on *Hypogymnia physodes* [140]. In particular, Pb concentration increases in a linear fashion as altitude increases, while Cd increases in the same way up to altitudes of 900–1100 m. For higher altitudes, Cd concentrations follow a decreasing trend. What is more, *H. physodes* is one of the most suitable bioindicators in the study of the bioaccumulation of trace elements [141] in view of its high-tolerance capacities.

In general, the higher accumulation of heavy metals in the thallus found after the summer period, may be due to the increased hydration that results from autumn rainfall [91]. In Mediterranean climates, the trace element content in lichens as they are (unwashed), is strongly influenced by soil dust contamination [142]. Loppi *et al.* [143] in spite of high correlation levels of Al, Fe and Ti in *Parmelia sulcata* does not find any linear correlation for these elements with their concentration levels in the soil. This would lead to the supposition that contamination through dust is highly variable and probably depends upon the local characteristics of the sites under study.

Cd is considered to be particularly toxic for various lichen species [144, 145]. Concentration intervals of 1.26–5.05 and 1.56–6.40 µg g^{-1} have been found for *A. ciliaris* and *L. pulmonaria*, respectively. These values (considering average values) are considered to be close to the appearance of toxicity symptoms. Furthermore, Cd has a high negative correlation with protein and reducing sugar content [146].

Lichens from the *Usnea* species have been used to evaluate heavy metal deposition patterns in the Antarctic [147]. The activities carried out in the different scientific stations could be potential sources of pollution and contribute to the circulation of trace metals in this site.

The relationship between cationic concentrations in lichens, as shown for *Cladonia portentosa*, can be used as an index of acid precipitation. In particular, the K^+/Mg^{2+} ratio and the (extracellular) Mg^{2+}/(intracellular) Mg^{2+} in lichen apexes is strongly correlated to H^+ concentrations in precipitation. High concentrations of H^+ that are found in acid rain cause increases in extracellular Mg^{2+}. In general, the variation in Mg^{2+} concentration in lichens can be considered to be a good indicator of acid rainfall [3, 148, 149].

Acid-moisture depositions containing heavy metals can significantly reduce lichen survival in affected geographical areas. In lichens (*Bryoria fuscescens*) exposed to simulated acid rainfall containing two levels of Cu^{2+} and Ni^{2+} only or combined with acid rain (H_2SO_4) at pH 3 for 2 months in addition to environmental rainfall, it was observed that the alga and fungus components respond in different ways to pH levels and that they have a specific interaction that is correlated to the toxicity of the metals. In particular, the alga component is the more sensitive to acid rain and to the mix of heavy metals and, as a result, it has a higher quantity of degenerated cells, which causes significant changes in membrane permeability. Critical concentrations of heavy metals in alga thalli were > 50 μg g^{-1} for Cu and > 7 μg g^{-1} for Ni in the presence of acidity and > 20 μg g^{-1} for Ni in absence of acidity [150, 151].

Another recently developed field of application for biomonitoring with lichens is that of indoor pollution and in particular, the analysis of air particulates. Rossbach et al. [152] found a high ratio between the concentrations of Cr, Zn and Fe in air particulate samples taken from the filters of air conditioning systems in different hotels in different cities and in Usnea spp. samples found in the conditioned environments.

Table 2 reports some bibliographical data on heavy metal bioaccumulation on lichens.

For over 20 years, lichens have been used as bioindicators and/or biomonitors in environment quality evaluations for such industrial realities as iron foundries and fertiliser manufacturing plants [153–155]; steel works and iron foundries [156–158]; oil extraction plants [159]; sites contaminated by the petrochemical industry [160]; areas surrounding zinc foundries [161]; areas surrounding nickel foundries [162]; coal-fired power stations [163–165]; power stations in high-density industrial areas [166]; petrochemistry, chemistry, metallurgy and power stations [167]; smelting industry [115, 117]. Garty et al. [107], in a study of a heavy oil combustion plant in the south-west of Israel (Ashdod area), carried out using transplanted of fruticose epiphytic lichens (Ramalina duriaei) found high concentrations of S, V and Ni in thalli that were transplanted to the industrial area. These values can be correlated with environmental measurements of SO$_2$ and V made in the same region. Furthermore, the high V/Ni ratio in the lichens may be an indicator (tracer) of pollution in the area caused by heavy oil combustion plants. The same authors [108] found a high Pb, V, Ni, Zn and Cu bioaccumulation potential for Ramalina lacera in the same site.

Lichens are also used to study copper dust emissions from mines. Ramalina fastigiata has been used as a bioindicator of the impact of a coal mine in Portugal. The threshold concentration of intracellular Cu above which total inhibition of photochemical apparatus occurs, is approximately 2.0 μmol g^{-1} [168]. Neophuscelia pulla and Xanthoparmelia taractica were used to study the bioaccumulation of heavy metals in abandoned copper mines in Greece, where a significant correlation ($p < 0.05$) was found between the copper content in the soil and that of the lichen thalli [169].

The aquatic lichen Dermatocarpon luridum was used as bioindicator of copper pollution [170]. In this work lichen thalli were exposed to 0.00, 0.25, 0.50, 0.75 and 1.00 mM copper in synthetic freshwater to solve the problems of metal bioavailability. The copper concentration extracted from thalli was correlated with pollution intensity and the obtained malondialdehyde (MDA) concentration in thalli can be used as indicator of copper pollution [170]. However, recent studies conducted on Parmelia species [171], indicate that copper toxicity is not a simple function of the Cu^{2+}-binding properties of the lichen substances present in the thallus.

Table 2: Selected references of heavy metals ($\mu g\ g^{-1}$ dry weight) studied on lichens species from different geographical areas.

Species	Site	Note	Cd	Cr	Cu	Fe	Mn	Ni	Pb	Zn	Ref.
Anaptychia ciliaris (N)	Southern Greece	Area of 250 km² of Peloponnesus where a lignite-burning power plant is located (RM)	3,09 1,26– 5,05		4,06 1,10– 5,60	2153 1359– 3092	43,87 15,6– 92,1		8,60 3,57– 12,6	31,22 23,3– 41,9	[146]
Lobaria pulmonaria (N)	Southern Greece	Area of 250 km² of Peloponnesus where a lignite-burning power plant is located (RM)	3,42 1,56– 6,4		6,85 4,6– 12,3	1103 339– 2180	65,32 17,7– 137,1		9,76 3,87– 21,1	28,16 16,9– 59,4	[146]
Ramalina farinacea (N)	Southern Greece	Area of 250 km² of Peloponnesus where a lignite-burning power plant is located (RM)	3,80 2,18– 7,06		3,63 1,70– 5,80	748 409– 1222	52,35 28,7– 81,4		11,18 5,10– 19,5	19,46 15,8– 25,6	[146]
Ramalina duriaei (N)	Israel	SC			5,5						[33]
Ramalina duriaei (T)	Israel	AC			12						[33]

Species	Location	Description									Ref.
Ramalina duriaei (T)	Israel	Urban (U), rural (R) and suburban (SU) sites that are 4, 5–24, and 5 km from a power plant		(U) 11,7 (R) 12,8 (SU) 23,8	(U) 13,6 (R) 12,9 (SU) 14,2			(U) 49,5 (R) 14,5 (SU) 20,0	(U) 165,4 (R) 30,3 (SU) 43,4	(U) 59,6 (R) 49,4 (SU) 59,0	[303]
Ramalina duriaei (N)	HaZorea (Northeast Israel)	SC	10.8	10			20.1	22.6		33.1	[303]
Ramalina duriaei (N)	Israel	Urban industrial area / Suburban area / Rural area					14.3 / 15.3 / 9.9		23.1 / 31.7 / 6.4		[107]
Ramalina lacera (T)	Israel	Southwest (AC)	6.53 (5.4–8.5)	7.34 (5.2–12.4)	1505 (1235–1954)	32.51 (26.4–40.9)	6.84 (2.1–12.2)	34.95 (10.4–155)	14.1 (6.7–26.2)	60.1 (38–113)	[108]
Evernia prunastri (T)	Rome	Area of 300 km² of the big annular connection (AC)	14.05	13.37	833.5	18		40	5.2	57.45	[134]
Parmelia sulcata (T)	Bern (Switzerland)	Urban area	28.5	47.5			213	189		259	[304]
Parmelia sulcata (T)	Biel, champagne Allee (Switzerland)	Urban area	22	84	890	21	22	172		191	[304]

Continued

Table 2: (Continued)

Species	Site	Note	Cd	Cr	Cu	Fe	Mn	Ni	Pb	Zn	Ref.
Parmelia sulcata (T)	Lauenen, (near Gstaad, Switzer-land)	Suburban area	526	18	4613	48	667	29		192	[304]
Parmelia sulcata (N)	Tuscany (provinces of Siena and Grosseto) (RM)	SC	3.6	9.1	1800	38.2		15.9	2.32	65.7	[143]
Parmelia sulcata (N)	Portugal	228 sites along the Atlantic coast and the interior of the country	5.77 1.53– 32.3				3.92 0.52– 33	18.4 2.0– 142	14.5 1.83– 130		[110]
Hypogymnnia physodes (N)	Village of Gusum (Sweden)	Surroundings of a brass foundry		28.2 11– 79	832 290– 1300		2.6 1.7– 3.9	22.6 14– 33		232 93– 450	[305]
Hypogymnnia physodes (T)	Bern (Switzer-land)	Urban area	n.d.	23	864	n.d.	30	315		224	[304]
Hypogymnnia physodes (T)	Biel (Switzer-land)	Urban area	30	41	1311	13	12	111		159	[304]

Species	Location	Notes									Ref.
Hypogymnia physodes (T)	Slovenia	The sites were located at least 300 m away from main roads and at least 100–200 m away from dwelling (RM)	5.78 (2.33–21.8)		1253 (492–3756)					90.2 (47.3–151)	[141]
Parmelia caperata (T)	Washington (near to the Potomac River)	The locations are near the Dickerson power plant	3.8 (2.0–5.1)		1400 (750–2090)	240 (140–380)				64 (55–80)	[163]
Parmelia caperata (T)	Travale-Radicondoli (Central Italy)	Area of 15 km² (Near a geothermal power plant (RM)	4.51 (1.25–8.41)	10.8 (4.5–25.4)	1019 (275–2370)	85.8 (10.9–280)	4.41 (1.65–8.18)	6.3 (2.1–19.7)		43 (22.2–63.8)	[182]
Parmelia caperata (N)	Former mining district (Central Italy)	(RM)	0.26 (0.06–0.69)	2.48 (1.19–5.66)	5.77 (3.94–9.17)	541 (161–2503)	65.5 (18.8–170)	2.65 (1.03–8.00)	3.88 (0.68–11.20)	34.7 (25.9–57.7)	[308]
Parmelia rudecta (T)	U.S.A. (near Potomac River Washington)	The locations are near the Dickerson power plant	4.6 (2.8–73)		1620 (780–3090)	230 (86–365)				68 (34–100)	[163]
Cetraria cucullata (N)	Northwest of Canada				478.8	47.7	2.5	4.2	1.78	24.1	[309]
Cetraria cucullata (N)	Taimyr Peninsula, Russia	(RM)	0.12 (0.066–0.382)		1.52 (1.12–12.79)	519 (290–1000)	34.3 (8.45–134)	1.52 (0.83–10.20)	1.26 (0.78–5.78)	20.5 (9.73–29.6)	[310]

Continued

Table 2: *(Continued)*

Species	Site	Note	Cd	Cr	Cu	Fe	Mn	Ni	Pb	Zn	Ref.
Cetraria nivalis (N)	Northwest of Canada				257.7	84.5	2.7	5.6	1.2	25	[309]
Cladina stellaris (N)	Northwest of Canada				568.7	30.2	2.9	4.3	3.98	15.9	[309]
Cetraria islandica (N)	Switzerland (Devos)	Rural area	8	8	149	43	7	40		60	[304]
Usnea filipendula	Slovenia	Surroundings of a brass foundry		22.4 7–48	614 250–1400		2.6 1.7–4.1	27 15–40		182 82–200	[305]
Usnea spp. (T)	Switzerland (Devos)	Rural area	36	14	402.5	26	32	120		72.5	[304]
Pseudevernia furfuracea (N)	Slovenia	Surroundings of a brass foundry		35 15–71	926 504–1640		2.8 1.9–3.9	37.3 20–53		237 125–444	[305]
Pseudevernia furfuracea (T)	Switzerland		41.8	34.7	3423	44.8	74.5	135.5		159.5	[304]
Cladonia rangiferina (N)	Slovenia	Surroundings of a brass foundry		14.5 4–40	442 160–1200		1.5 0.8–2.3	22.8 11–36		102 51–204	[305]
Umbilicaria deusta (N)	Sudbury District (Northern Ontario, Canada)	Surroundings of a copper foundry		65	1470		37				[306]

Umbilicaria muhlenbergii (N)	Sudbury District (Northern Ontario, Canada)	Surroundings of a copper foundry	30	920	16			[306]
Stereocaulon paschale (N)	Sudbury District (Northern Ontario, Canada)	Surroundings of a copper foundry	43		26			[306]
Cladonia alpestris (N)	Mackenzie Valley (Ontario, Canada)	Boreal forest	12.6	164				[306]
Cladonia alpestris (N)	Sudbury District (Northern Ontario, Canada)	Surroundings of a copper foundry	15	320	11			[306]
Cladonia deformis (N)	Mackenzie Valley (Ontario, Canada)	Boreal forest	1	70				[306]
Cladonia deformis (N)	Sudbury District (Northern Ontario, Canada)	Surroundings of a copper foundry	21	220	10			[306]

Continued

Table 2: (Continued)

Species	Site	Note	Cd	Cr	Cu	Fe	Mn	Ni	Pb	Zn	Ref.
Cladonia mitis (N)	Mackenzie Valley (Ontario, Canada)	Boreal forest		20.6	170						[306]
Cladonia mitis (N)	distretto di Sudbury (Ontario del Nord, Canada)	Surroundings of a copper foundry		19	260		10				[306]
Cladonia uncialis (N)	distretto di Sudbury (Ontario del Nord, Canada)	Surroundings of a copper foundry		19	210		10				[306]
Cladonia uncialis (N)	Mackenzie Valley (Ontario, Canada)	Boreal forest		4	120						[306]
Diploschistes steppicus (N)	Sede Boquer region (Negev desert, Israel)	In this area no industry exists, we may assume that the heavy metals derive partly from car traffic	17.2	66.9		66	23.6	63		33.2	[307]

Species	Location	Notes							
Teloschistes lacunosus (N)	Sede Boquer region (Negev desert, Israel)	In this area no industry exists, we may assume that the heavy metals derive partly from car traffic	15.2	4.7	33.5	12.8	19.4	30.3	[307]
Quamarina crassa (N)	Sede Boquer region (Negev desert, Israel)	In this area no industry exists, we may assume that the heavy metals derive partly from car traffic	18	14.4	68.6	21.2	31.5	41.9	[307]
Ramalina maciformis (N)	Sede Boquer region (Negev desert, Israel)	In this area no industry exists, we may assume that the heavy metals derive partly from car traffic	10.3	8.1	8.1	11.9	39.6	20.9	[307]
Caloplaca ehrenbergii (N)	Sede Boquer region (Negev desert, Israel)	In this area no industry exists, we may assume that the heavy metals derive partly from car traffic	20	8.7	42.4	14.4	35.3	28.8	[307]

Mean values and ranges of concentrations; T = transplanted, N = natives, AC = contaminated area, SC = control site, n.d. = not detectable, RM = reference material used (modified from Conti and Cecchetti [3]).

Hypogymnia physodes has been used as a bioindicator of the presence of mercury and methylmercury in metal extraction areas in a site in Slovenia, where its excellent bioaccumulation capacities were confirmed [172]. *Parmelia sulcata* was also employed to evaluate mercury pollution near a thermometer factory [173]. Levels of about 0.2 mg kg^{-1} were found in lichens collected about 20 km away from the factory, near a pristine lake area. However, very few methods have been reported for the speciation of mercury in lichens [172]. The development of techniques for the separation and identification of individual mercury species in environmental samples remains a critical issue [174]. As for instance, Balarama Krishna *et al.* [175], proposed a rapid ultrasound-assisted thiourea extraction method for the analysis of total mercury, inorganic and methylmercury (MM) in some environmental samples such as lichens. Results showed close agreement with certified values with an overall precision of 5–15%. Quantitative recovery of total Hg for lichens was achieved using a mixture of 10% HNO_3 and 0.02% thiourea.

The sorption and desorption behaviour of inorganic, methyl and elemental mercury on *Parmelia sulcata* and moss (*Funaria hygrometrica*) samples, studied under laboratory conditions, suggest that the lichens and mosses can also be used as sorbent material for the decontamination of inorganic and MM from aqueous solutions [176].

Riga-Karandinos and Karandinos [146] took three native lichen species (*Anaptychia ciliaris*, *Lobaria pulmonaria* and *Ramalina farinacea*) directly from 22 sites distributed over an area of 250 km^2 in Peloponnesus, Greece, where there is a coal-fired power station. There they found levels of Cd that were close to levels of toxicity for lichens (respectively 3.09, 3.42 and 3.80 µg g^{-1}) and below-toxic levels of Pb (respectively 8.60, 9.76 and 11.18 µg g^{-1}).

Lichens are excellent bioindicators of atmospheric pollution from geothermal power stations and in particular, of the pollutants that are correlated with this phenomenon, such as mercury [177], boron [178], radon [179] and other metals [180]. Several authors [25, 181] give data on atmospheric pollution from geothermal power stations in central Italy using the method of mapping lichen communities, where they found minimum IAP values within 500 m of the power station and progressive increases in frequency in line with increases in distance from the power stations themselves. Pollutants that are typically associated with geothermal activity are As, B, Hg and H_2S [25, 182]. Nonetheless, it is not clear if the drop in the richness of the species in those areas close to geothermal power stations is due to the action of a single contaminant or to the combined actions of all contaminants. In any case, the author considers that the worst damage to lichen thalli is caused by H_2S, which is a highly toxic gas [183] and which is continuously present in high levels as a local contaminant in areas surrounding geothermal power stations.

Different studies have established correlation factors between chlorophyll damage and the concentrations of several elements in lichens. Garty *et al.* [108] found that chlorophyll integrity is inversely correlated with concentrations of Cr, Fe, Mn, Ni, Pb and B. K concentrations, however, have a positive correlation with chlorophyll integrity [108, 153]. In a study of *Cladina stellaris* samples that were transplanted to an area near to a fertiliser factory in Finland, Kauppi [153] found that the high concentrations of K in the lichens corresponded to increases in chlorophyll content and in particular, to the Chl *a*/Chl *b* ratio.

Garty *et al.* [108], found differences of approximately factor 2 in electrical conductivity in lichens from industrially polluted sites (Israel) compared with those from rural sites. This indicates a process of cell membrane damage. Membrane integrity is highly correlated with the presence of Ca [145], which is a macronutrient and has regulatory functions in sites of extracellular interchange on the surface level of alga cell walls or hyphae, and intracellular interchanges with proteins. Concentrations of several elements (S, B, Al, Cr, Fe, Si, Ti and Zn) are positively correlated with cell membrane damage for *R. duriaei* [3, 40, 184].

Various studies report that exposure of lichen thalli to chemicals in laboratory and using fumigation experiments, reveals subsequent damage to cell membranes corresponding to an increase in water electrical conductivity caused by the loss of electrolytes [184–186]. In general, the loss of K is related to cell membrane damage and is inversely correlated with electrical conductivity [151, 184]. The extent of cell membrane integrity may be evaluated by measuring electrical conductivity [187]. Conductivity was mostly related to released Na, Cl, K, Mg and Cs in a survey conducted on *Parmelia sulcata* transplants [188]. The obtained concentrations suggested that Na, Cl and K could be considered as largely determining the conductivity. Generally, the comparability of lichen vitality in wide geographical areas seems to be of a limited value and controlled by the area's variability in temperature and precipitation rather than variability in metal deposition rates [189, 190].

Carreras *et al.* [191] pointed out that physiological mechanisms (i.e. malonaldehyde content and electrical conductivity) involved in metal uptake in lichens should be carefully analysed when studying heavy metal impact of an environment. The AA found possible competitive mechanisms of cation uptake (i.e. Pb and Cu) for *Usnea amblyoclada*. Pb seems to have higher affinity for the lichen cell wall exchange sites than the other metals; pollutants could be damage cellular membranes thus altering mechanisms of uptake of metals.

Another field of study using lichens as biomonitors is that involving volcanic areas [72] and in particular, the release of mercury. In the Hawaiian islands, Hg has been found in the interval of 8–59 µg g^{-1}, meaning that bioaccumulation is therefore more widely distributed in areas that are strongly affected by volcanic activity [192].

An important problem – as far as concerns the determination of trace elements – is the quality control of analytic methods as well as of the sampling strategies and treatment of samples [193]. In the last few years (2000–2007), more than 900 articles on the analysis of lichens have been published. The high variability of the data found may not only be caused by the different distribution of pollution patterns, but it may also be the result of possible errors in analysis. It is in this sense that the European Commission, through its Standards, Measurement and Testing programme has developed, some years ago, a CRM of lichen (CRM 482) to determine and remove the major sources of error in lichen analysis [194]. The International Atomic Energy Agency (IAEA) also developed, in an interlaboratory study with developing countries, a reference material IAEA 336 lichen [195].

Within the framework of the IAEA Analytical Quality Control Service (AQCS) many research programs were initiated for the identification of trace elements in candidate IAEA lichen-338 for a proficiency tests (PTs) [196, 197]. For Cd, Cu, Fe, Mn, Pb and Zn, analysed by electrothermal atomic absorption spectrometry (ETAAS) the results obtained in CRMs were in good agreement with the certified values and the recoveries were about 100% [196].

The homogeneity of the candidate CRMs (IAEA-338 Lichen, IAEA-413 Algae) were tested by using different analytical techniques such as neutron activation analysis (NAA), synchrotron radiation X-ray fluorescence and macroproton induced X-ray emission (macro-PIXE) [198, 199]. In general the two candidate RMs showed good homogeneity of the tested samples [198]. Also the particle size in the IAEA material (grain size <125 μm) determined by the algae/fungus ratio percentage could influence the elemental content in the IAEA sample [200].

It should be noted that the problem of sample pre-treatment and sampling methods, as well as the standardisation of analytical methods, is of fundamental importance above all when comparing lichens from different geographical areas [97, 201].

Literature contains little information above all about the techniques for washing lichens. Different lichen washing strategies may cause relevant changes in metal or sulphur content [202] or in fluorides [155], as against unwashed samples.

Furthermore, the enormous developments in analytical techniques (atomic absorption spectroscopy, Inductively Coupled Plasma-Atomic Emission Spectroscopy (ICP-AES), neutron activation, etc.) should be taken into consideration [203]. These developments have, in the last few decades, notably improved instrument detection limits, allowing increasingly accurate analyses and eliminating possible sources of error.

Novel digestion methods of samples were recently proposed [204]. For example, a simple and efficient digestion method for rapid sample preparation and quantification of 25 chemical elements in lichens by sector field inductively coupled plasma mass spectrometry (SF-ICP-MS) was developed [205]. The method of microwave-assisted acid digestion (MW) was carried out at atmospheric pressure simultaneously handling up to 80 samples in screw-capped disposable polystyrene tubes. This new procedure was then compared with the established MW digestion in closed vessels in order to examine its potential applicability in routine analysis for environmental monitoring. The majority of the elements were totally recovered from the lichens giving reliable results, low contamination risk, simplicity, timesaving and applicability in routine environmental analyses [205].

The application of ultrasound-assisted extraction procedure has been also examined for the estimation of major, minor and trace elements in lichen samples as a possible alternative to conventional digestion methods [206]. This study indicates that the method is a fast (within 15 min including centrifugation time) and simple method for the estimation of many elements such as Na, K, Ca, Mg, Cr, Mn, Co, Ni, Cu, Zn, Ge, As, Se, Rb, Sr, Zr, Ag, Cd, In, Sb, Cs, Ba, Pb and Bi [206].

Paul and Hauck [207] studied the effect of incubation with $MnCl_2$ on chlorophyll fluorescence parameters in lichens. Incubation with $MnCl_2$ decreased the effective quantum yield of photochemical energy conversion in photosystem 2 in *H. physodes* as well as in cyanolichens *Leptogium saturninum* and *Nephroma helveticum*, but not in *Lobaria pulmonaria*.

The sensitivity of the Chl fluorescence parameters to the incubation with $MnCl_2$ clearly differs between species. While *H. physodes*, *L. saturninum* and *N. helveticum* were differently sensitive to $MnCl_2$, none of the investigated parameters was changed by $MnCl_2$ in *L. pulmonaria*. This result is remarkable, as *L. pulmonaria* is very sensitive to SO_2 [3], but this low toxitolerance is apparently not applicable to Mn. This seems to imply that different mechanisms are responsible for Mn and SO_2 toxicity symptoms in lichens [207].

Salemaa *et al.* [208] studied macronutrients, heavy metals and Al concentrations in bryophytes, lichens and vascular plants species growing in Scots pine forests at four distances from the Cu–Ni smelter plant and then they were compared to those at two background sites in Finland.

The authors studied the relationship between element accumulation and the distribution of the species along a pollution gradient. Elevated sulphur, nitrogen and heavy metal concentrations were found in all species groups near the pollution source. Macronutrient concentrations tended to decrease in the order: vascular plants > bryophytes > lichens, when all the species groups grew on the same plot. Heavy metal concentrations (except Mn) were in the order: bryophytes > lichens > vascular plants.

When determining trace elements in lichens, some interesting analytical applications, such as the use of short-life radionuclides in neutron activation [209] and X-ray fluorescence spectrometry [210, 211] can supply valid information, in particular for the determination of macronutrients (K and C), or of trace metals (Cu, Pb, Zn) and non-metals (S).

2.4   Sulphur compounds

The effect of sulphur compounds on lichens have been extensively studied (i.e. on urban lichen diversity and vitality) [3, 212–214]. For indeed, many studies are generally concerned with the effects of SO_2 fumigation of exposed lichens [8, 36, 215–217], or with the effects of simulated acid rainfall [150, 218–221]. Other works deal in the respiration rate [222], photosynthesis [153, 223, 224] and chlorophyll fluorescence [32, 225, 226]. For the most part, these studies aim to evaluate the effects of sulphur compounds on the physiology of lichen thalli and/or on the integrity of photobiont chlorophyll.

Chlorophyll analysis is usually carried out following the method proposed by Ronen and Galun [47]. Phaeophytin (Ph) is a product of chlorophyll degradation. The variation in the normal *chlorophyll–phaeophytin* ratio is an indication of suffering in lichens. This method foresees the extraction of chlorophyll using 5 ml of solvent (DMSO – dimethyl sulfoxide technique). The ratio between Chl a and Ph a is measured using a spectrophotometer [Optical density (OD) 435 nm/OD 415 nm]; and this is considered to be an appropriate index for measuring the impact of high concentrations of SO_2 in lichens, or for evaluating the effects of heavy metal pollution in transplanted lichens [164, 227]. A ratio of 1.4 indicates that chlorophyll is unchanged. Any reduction in this value indicates chlorophyll degradation with ensuing stress to the organism [43, 166, 228–232]. Kardish *et al.* [34] report a value of 1.44 for the Chl/Ph ratio of *Ramalina duriaei* in the control site, while for a polluted site with high levels of vehicle traffic; they found a value of 0.80. In general, an alteration of the Chl/Ph ratio has been found, indicating the toxic effect of a combination of gaseous and non-gaseous pollutants.

Sulphur content is determined by transforming elementary sulphur into SO_4^{2-} ions, which occurs through the acid suspension method using barium chloride [229] (expressed in mg g^{-1} dry weight). New developed methods such as continuous flow-isotope ratio mass spectrometry (CF-IRMS) allows direct measurement in lichens of sulphur isotopic composition while requiring little or no time-consuming chemical pre-treatment of the samples (at a level of µg S in lichen) [233].

Some authors [46, 230, 231] report that data relating to sulphur accumulation and obtained indirectly from the bioindicator, seem to show that the influence of SO_2 from industry (Córdoba, Argentina), is rather restricted compared to that which comes from vehicle traffic.

Sensitivity to SO_2 and to other atmospheric pollutants in general, varies according to species [234]. *Lobaria pulmonaria* is considered to be one of the most sensitive species according to the scale of [235]: 30 µg m^{-3} for average winter concentrations of SO_2. This species' high degree of sensitivity is probably due to the presence of isidia, which is a vegetative structure on the upper surface of the thallus and which plays a role in asexual reproduction. The isidia increases the absorption surface of the thallus per unit of mass [146]. *Hypogymnia physodes* is, on the contrary, a species that is particularly resistant to SO_2. Indeed, it has been seen that exposure of this species to H_2SO_4 in highly acid conditions, produces no effect [236]. *H. physodes* has also been used in the area surrounding a fertiliser plant in Finland, where sulphur levels of 3000 ppm had been found [155]. Typical ultrastructural damage caused by the action of sulphur on photobiont cells is seen within the first 2 weeks of the transplant, without the sulphur concentration levels being particularly high.

Another interesting field of research is that which correlates sulphur content and the composition of sulphur isotopes [202, 237–239]. A study by Wadleigh and Blake [202] reports the spatial variation of the sulphur isotope composition of 83 epiphytic lichen samples (*Alectoria sarmentosa*). The study reveals a positive correlation between isotope composition and different sources of sulphur emission in the site under study (Newfoundland Island, Canada). It is interesting to note the fact that lichens are also sensitive to the sulphur salts that come from the sea. Indeed the degree of sulphur concentration has been seen to decrease in lichens, the further they are from the coastal to internal areas of the island.

More recently, Wadleigh [240] transplanted thalli of *A. sarmentosa* from a relatively pristine area to an area of higher anthropogenic S input and monitored for 1 year. The transplanted lichens gradually approached the S concentration and isotopic character of the new site. Isotopic values (δ^{34}S) measured on old and young portions of several lichen thalli collected near an oil refinery showed that the youngest parts resembled the refinery emissions, while the oldest parts more closely matched the natural background signature [240]. Wiseman and Wadleigh [241] also established that 18 months are necessary for transplanted lichens (*A. sarmentosa*) from a remote area to an urban area, to re-equilibrate to a changing atmosphere in terms of sulphur concentration and isotopic composition.

The stable isotope abundance of carbon (δ^{13}) in the lichen *Cladia aggregata* has been investigated [242] in a study relating these values with known levels of SO_2, NO_x and O_3 in the region within a 200-km radius of Sydney, Australia. Small effects on δ^{13}C values in the lichen were found to be associated with distance from pollution source, humidity, altitude and number of rain days at the sampling site. A significant negative correlation between δ^{13}C values in the whole lichen and atmospheric SO_2 concentrations was observed. These results do not fit well to that observed in laboratory studies by other workers. More research is needed to clarify these relationships; a hypothesis of these differences could be attributed to the selected lichen species and their specific interaction with pollutants and δ^{13}C isotopes [242].

The role of sugars in alga–fungus interaction is most important in lichen biology. Chronic SO_2 fumigation of lichens may cause interference in the flow of such nutrients as carbohydrates, thus creating symbiont damage. SO_2 causes reducing sugars to increase and non-reducing sugars to decrease. This effect is probably due to a breakdown in the polysaccharides that are rich in reducing sugars. Reducing sugars are determined by extracting 10 mg of lichen thallus with 1 ml of d-H_2O and centrifugation at 2000 $\times$ g for 10 min in an Eppendorf vials. 4 ml of a mix of two solutions (sodium potassium tartrate and an indicator) are added to the supernatant and the vials were bathed in darkness at 100 °C for 3 min. After the proper cooling period, solution absorbance is measured at 660 nm [146].

Spectroscopic measurements, carried out to study changes in the spectral reflectance response of lichen thalli have been exposed in contaminated sites as against those exposed in control sites [107, 243]. As a rule, lichen scanning revealed, as for the higher plants of uncontaminated sites, a significant drop of between 600 and 700 nm (which corresponds to the absorption region of the chlorophyll), and a net increase in spectral reflectance around 700 nm (red edge) together with a continual and relatively high reflectance in the near infrared (NIR) between 700 and 1100 nm. The NIR plateau is the result of differences between the varying refractive indexes of the internal components of the thallus (cell walls, chloroplast, air, water content, etc.). In spectra of lichens transplanted to contaminated sites, the red edge (700 nm) is much less pronounced and the plateau is very low. This indicates a clear situation of organism stress [107]. In extreme cases, for example in plants that are subjected to high stress levels, or which are already dead, the spectrum shows a continuous line that rises gradually.

Membrane proteins may be damaged by the presence of SO_2, which may cause a reduction in protein biosynthesis in some lichens; or there may be negative effects on the nutritional interchange between symbionts with, as a consequence, an alteration of their delicate balance. To determine protein content, 100 mg of tissue are extracted using 3 ml of a phosphorous buffer solution at pH 7 and the extract is centrifuged at 1600 × g for 5 min. 100 µl of the extracted solution are added to 5 ml of Bradford solution [244] and absorbance at 595 nm is taken and compared with a calibration curve made following with protein standard (e.g. bovine serum albumin-BSA) [146]. The structural proteins found in cell membranes and the lichens enzymes can have considerable damage in the presence of high levels of SO_2 concentration. These processes concern the delicate interchange of nutrients between symbionts and can damage the delicate equilibrium of the association [41].

Thus, the damage to cell membranes can be used as an indicator of environmental stress. Indeed, it has been demonstrated that SO_2, such as O_3 and NO_2, are powerful catalysts of lipid membrane peroxidation [229, 231, 245]. Experiments where lichens were exposed to 1 ppm of SO_2 in aqueous solution show a slight reduction in the overall content of phospholipids and an increase in unsaturated fatty acids. This latter type of response to SO_2 may be considered to be of the adaptive type [246].

The effect of SO_2 can also be evaluated by dry weight/fresh weight ratio (DW/FW). This ratio has been proposed as an indication of the influence of the environment on the bioindicator. It has indeed been observed that in highly polluted areas (e.g. where traffic is intense), there is a tendency in lichens to lose moisture [230].

Finally, the production of ethylene is another indicator of stress in lichens [247]. Lichens exposed to solutions containing sulphur in an acid environment have different levels of ethylene production. In general, these solutions increase the solubility of the particles containing heavy metals that are trapped within the hyphae. This phenomenon may lead to an increase in the production of endogenous ethylene in lichens when they are exposed to chemical agents containing sulphur, to acid rain and to air polluted with heavy metals [236].

2.5 Nitrogen and phosphorous compounds

Although lichens have already been proposed as bioindicators of NH_3 [248], only in the last few years has a clear positive correlation been established between nitrophytic lichens and atmospheric NH_3 concentrations, even if responses are always greater to SO_2. Tree bark analyses in sites in Holland demonstrate that nitrophytic lichen species do not respond directly to nitrogen levels found in the environment, but that they are favoured by the high pH values in the bark, which are related to the high levels of NH_3 in the environment [249].

Cladonia portentosa is an excellent bioindicator for the study of precipitation chemicals and nitrogen and phosphorous concentrations. Hyvarinen and Crittenden [250, 251] have found concentrations in the range of 0.08–1.82% for nitrogen and 0.04–0.17% for phosphorous (per unit of dry weight) in apexes (5 mm top part) and thalli (bases) in various comparison sites. The concentration levels found for these elements are 2–5 times higher in the apexes than at the bases and furthermore, both the apexes and base parts show a high positive correlation between elements. The correlation between N deposition and the nitrogen accumulated in the lichens is positive; becoming higher when referred to concentrations found in the thalli rather than in the apexes. On the other hand, nitrogen concentrations in the thallus are little correlated with the N values in precipitation. The nitrogen found in thalli is, however, highly linked to moist nitrogen deposits, but it is also correlated positively with the NO_2 present in the air. As well as *C. portentosa*, *H. physodes* has also been proposed as a bioindicator of nitrogen total deposition (dry and wet) [252] as well as of nitrogen and sulphur [253].

High levels of SO_2 and NO_x can cause a reduction of pH of lichen thalli (see for example data respective to Peloponnesus area (Greece) [146]). To this respect, it shall be highlighted that atmospheric pollution of this kind has led to extinction of *L. pulmonaria* and *R. farinacea*. The measurement of the pH of lichen thalli can supply information with regard to the state of pollution of a site. To determine pH levels, 50 mg of lichen thallus is homogenised in liquid nitrogen and 2 ml of d-H_2O. After centrifugation at $100 \times g$ for 10 min, the pH value for the supernatant is read. Various authors report that *L. pulmonaria* is endangered in some sites subject to acid rain, and pH = 5 has been indicated as a threshold value below which lichen is unable to survive [3, 254, 255].

Russow [256] proposed a novel natural ^{15}N method that uses the lichen *Squamarina* with symbiotic green algae (which are unable to fix N_2) as a reference in order to determine N_2 fixation by biological soil crust of the Negev Desert. The relative biological fixation of atmospheric nitrogen was estimated at 84–91% of the total N content of the biological soil crust. The cyanobacteria-containing soil lichen *Collema* had a fixation rate of about 88%. These fixation rates were used to derive an absolute atmospheric N input of 10–41 kg N ha^{-1} year^{-1}.

Lichens and tobacco plants, were recently used in Grenoble (France) [257] to estimate, respectively, the spatial distribution of NO_x and O_3. Bioaccumulation of nitrogen in *Physcia adscendens* and bioindication of ozone as shown by visible injury to tobacco plants (cv. Bel W3) were studied. Results showed complementary biomonitoring maps, the first based on nitrogen concentrations of lichens and the second based on tobacco leaf injuries. The maps were highly comparable with results obtained from physicochemical analyses and then giving a realistic picture of the spatial distribution of atmospheric nitrogen oxides and ozone [257].

2.6   Ozone

O_3 and NO_2 (see also Section 2.4) are powerful catalysts of lipid membrane peroxidation: the main effect of O_3 on lichens is indeed the damage of cell membrane.

It has been demonstrated that in biological systems the presence of oxidation products such as MDA is directly correlated to the start of the peroxidation of unsaturated fatty acids [258]. Egger *et al.* [259] reports an increased production of both MDA and superoxide dismutase in *Hypogymnia physodes* that was transplanted to highly polluted sites with monthly O_3 concentrations in the range of 20–198 μg m^{-3} (10–100 ppb). These compounds are products of lipid peroxidation and are indicators of oxidative damage to membranes and to the enzyme systems that protect against oxidation in plants [33].

Oxidation products are estimated by determining MDA, which is measured using the colorimetric method [260]. MDA is determined using the extinction coefficient of 155 mM^{-1} cm^{-1} [261]. The results are expressed in μmol g^{-1} of dry weight.

Other important peroxidation products are hydroperoxy conjugated dienes (HPCD) which can be isolated by solvent extraction. Concentration is calculated using the molar extinction coefficient $\varepsilon = 2.64 \times 10^4$ M^{-1} m^{-1}. Results are expressed in mmol g^{-1} of dry weight.

O_3 damage to the photochemical apparatus of lichens after repeated exposure to real doses has been well documented [262]. In particular, this kind of damage has been studied in other species such as, for example, clover, where pollution causes typical and easily recognisable leaf damage [263]. Furthermore, O_3 is the subject of monitoring and controls made by the European Community [264]. Ross *et al.* [265], report a study where *Flavoparmelia caperata* was fumigated with O_3 for brief periods (10 and 12 h) and in quantities of 200 μg m^{-3} (100 ppb). This caused a 50% net decrease in photosynthesis. *Usnea ceratina* was fumigated for 6 h per day for a period of 5 days with concentrations of 100–200 ppb of O_3, causing a notable reduction in net photosynthesis [49].

2.7 Fluorides, chlorides and other atmospheric pollutants

Literature regarding the bioaccumulation of fluorides is scarce. Asta and Garrec [266] have demonstrated that fluoride concentration in the lichen thallus is dependent on both the lichen species and the environmental F levels. Fluoride levels above 360 ppm were found in lichens (*H. physodes* and *Bryoria capillaris*) transplanted to areas near to a fertiliser factory and mine [155]. Fluoride accumulation in these sites reached maximum levels during the summer followed by decreases in the autumn. Already at levels of 30–40 ppm of dry weight it is possible to see the typical ultrastructural damage in the photobiont cells, caused by exposure to fluorides. It has also been found that fluoride content in lichens is inversely correlated to the distance from an aluminium processing plant. The losses in the lichen populations around this plant show a high level of correlation ($r^2 = 0.90$) with their F content [267].

Studies inherent to the impact of chlorides are also rather scarce. To this regard, a biomonitoring study using the lichen *Parmelia sulcata* Tayl. and several mosses, carried out in 26 sites, shows a spatial and temporal correlation between chloride bioaccumulation and the environmental impact trend of a waste incineration plant in the city of Grenoble, France [268].

For other atmospheric contaminants, such as polychlorinated dibenzodioxins and polychlorinated dibenzofurans (PCDD$_s$/PCDF$_s$), bioindication studies using lichens are very scarce. In general, the available studies of PCDDs and PCDFs report the temporal variations of these pollutants in samples of different types (e.g. plants) taken in the vicinity of urban waste incineration plants or other industrial areas [269–271]. *Xanthoria parietina* seems to be a good biomonitor for monitor PCDD$_s$/PCDF$_s$ deposition [271]. Furthermore, for example, the eggs of several bird species and in particular, those of the herring gull, seem to be good bioindicators of the presence of PCDD$_s$/ PCDF$_s$ [272].

More recently, Domeño *et al.* [273] developed a fast method for the extraction of polycyclic aromatic hydrocarbons (PAHs) by using the dynamic sonication-assisted solvent extraction from *Xanthoria parietina* samples. Sixteen PAHs were extracted from 0.2 g of dried lichen, in 10 min with a total extraction volume of 2 ml of hexane. The preconcentrated extracted fraction was analysed by GC–MS without any clean-up step following extraction. Both spiked and non-spiked native samples were used for the evaluation. The procedure was compared with the static ultrasonic and Soxhlet extraction and showed high efficiency with respect to both recoveries, time and solvent consumption and for all the 16 PAHs recoveries higher than 70% were obtained [273].

Blasco *et al.* [274] demonstrated that *Parmelia sulcata* is a potential useful bioindicator of atmospheric pollution by PAHs. Samples of *P. sulcata* were collected from sites on both sides of the Somport tunnel (which links France and Spain) and atmospheric particles were collected by air samplers installed within and on either side of the tunnel. The PAH concentrations in the particles were found to be nearly 2 orders of magnitude higher than the concentrations in the lichens. The total concentration of 16 priority studied PAHs were in the range from 6.79 to 23.3 $\mu g\,g^{-1}$ in particles outside the tunnel, from 18.3 to 265.2 $\mu g\,g^{-1}$ in particles inside the tunnel, and from 0.91 to 1.92 $\mu g\,g^{-1}$ in the lichen samples. The obtained results from the lichen samples seem to confirm that they may be excellent biomonitors of PAHs pollution in remote areas. The main source of the PAHs found in the lichens and particles seems to be road traffic, but other sources can have made minor contributions.

Favero-Longo *et al.* [275] studied the interaction between some lichen species and substrata bearing fibres of chrysotile [$Mg_3Si_2O_5(OH)_4$], that is the most widespread commercial form of asbestos. A natural deactivation of chrysotile asbestos occurs on serpentinite rocks where lichens (both crustose [*Candelariella vitellina* (L.) Müll. Arg. and *Lecanora rupicola* (L.) Zahlbr.] and foliose [*Neofuscelia pulla* (Ach.) Essl. and *Xanthoparmelia tinctina* (Maheu & A. Gillet) Hale]) selectively grow on the fibres and secrete metabolites such as oxalic acid, which, in the long term, turn the fibres into a non-toxic amorphous material.

2.8   Radionuclides

Lichens are good bioaccumulators of radionuclides [276]. This application concerns an important sector of research into the evaluation of the fallout of radionuclides, above all after the Chernobyl incident [277–285]. In lichens in several areas of Norway, after the Chernobyl incident of 1986, levels of Cs 134 and Cs 137 of two orders of size larger than those in vascular plants were found [286]. This phenomenon has caused a significant increase in the average concentration of radiocesium in reindeers, whose major food source are lichens [287, 288]. *Parmelia sulcata* has been used as a bioindicator for the presence of radionuclides in areas close to Chernobyl where I 129 and Cl 36 have been measured. Regional distribution patterns of these radionuclides have shown a positive correlation with accumulated concentrations [289]. One study reports values of the natural decontamination of radionuclides in lichens [290]. The average biological life span of Cs 137 is 58.6 months in *Xanthoria parietina*, which has shown it to be the best bioindicator of radioactive fallout as against mosses.

It has been demonstrated that plutonium concentrations in *Xanthoria spp.* gathered in the vicinity of a nuclear arms deposit are inversely correlated with the distance from the contamination site, and that there is a direct correlation between concentrations of Pu 239 and Pu 240 with concentrations found on the soil surface ($r^2 = 0.767$; $p < 0.001$) [291]. Altitude is an important factor that is correlated with concentrations of Ra 226 and Ra 228 studied in lichens of the *Umbilicaria* species [292]. Altitude is also correlated with levels of Cs 137 found in 1993 in samples taken in Italy in the province of Parma [283].

Hypogymnia physodes was used as bioindicator of uranium content in lichens and air samples as well as isotopic mass ratios $^{235}U/^{238}U$ in the central region of Russia from 1999 through 2001 [293]. Two sites were selected for this study: one was near of an uranium emission source ("contaminated") and the other at the distance of 30 km in the predominantly upwind direction. The mean measured content of U were 1.45 mg kg^{-1} in lichen at 2.09×10^{-4} µ m^{-3} in air in the "contaminated" site, whereas in the "clean" site the U levels were 0.106 mg kg^{-1} in lichen at 1.13×10^{-5} µ m^{-3} in air [293]. The authors also estimated the uranium content and its isotopic ratio in lichens taken from the "clean" site. Results were 0.062 mg kg^{-1} d.w. in lichen sample (in the extract of 1 M of HNO$_3$ from ash) and 0.0050 for $^{235}U/^{238}U$ isotope mass ratio. In the insoluble ash residue of lichen the U content was 0.029 mg kg^{-1} d.w. and 0.0072 for $^{235}U/^{238}U$ isotope mass ratio. Rosamilia *et al.* [294] reported data obtained by the UNEP field study on uranium isotopes and metals in lichens and tree barks. This study was performed in Bosnia-Herzegovina, where depleted uranium (DU) ammunition was used by NATO during the Balkans conflict (1994–1995). The results of this survey showed that the presence of DU in lichen and tree/bush bark is due to a direct deposition of DU dust particles during the attack and the deposition of suspended materials in air (originating from resuspension of soil and deposited DU dust particles).

Comparison of cosmogenic ^{7}Be activity concentrations measured in lichens with different atmospheric deposition rates confirmed that *P. sulcata* can be used as a quantitative biomonitor of radioactive trace substances [285].

Lichens collected in France in the surroundings of a military nuclear facility, a reprocessing plant and in a reference site were analysed for organically bound tritium (OBT) and radiocarbon contamination. Results showed in the most contaminated sites levels of OBT higher than the background levels by a factor of 1000. Tritium and radiocarbon are incorporated by photosynthesis, the slow metabolism of lichens makes them suitable for the follow-up of ^{3}H and ^{14}C [295, 296].

3   Concluding comments

The analysis of atmospheric pollutants using conventional analytical procedures allows data to be interpreted directly and results to be obtained rapidly. A summary of the most common conventional analytical techniques for the analysis of various environmental pollutants is given in table 1 (Chapter 1).

On an ecological level, however, air quality studies using these methods may present the following problems:

1. space-time fluctuations could lead to sampling errors;
2. low concentrations of several microcontaminants (which could also change over time), could lead to difficulties in methodology;
3. it is also difficult to ascertain either the intermittent or sporadic emission of contaminants;
4. in this way the biological tolerance limits of the species concerned might not be taken into consideration;
5. often the dose-effect ratio does not have linear response and it is thus possible to run into interpretation problems in evaluating damage to organisms and ecosystems.

The above points highlight the fact that traditional environmental monitoring methods require numerous and extensive samples to be taken in the areas under study and that these samples must also be taken over prolonged periods of time.

Furthermore, the use of mathematical models of contaminant dispersion in the environment should also be pointed out [297]. These models, based on physical and chemical properties, have produced excellent results in the last few years, especially as far as predictions of contaminant dispersion and potential bioavailability are concerned. However, these methods are little developed, above all with regard to effects on species and ecosystems [298].

These models of pollutant propagation and transport usually, however, concern punctual sources of contamination and they also require large quantities of information if they are to be applied. From this stems the importance of bio-monitoring when establishing contaminant levels in organisms and to use in ascertaining possible toxicity in relation to organism placement within the ecosystem.

From an ecotoxicological viewpoint, it is very difficult, starting from chemical analysis, to establish a valid model for toxicity capable to foresee the bioavail-ability and various complex synergies that run between the organisms present in an ecosystem. For instance, some efforts were recently conducted in order to develop models for the assessment of metal bioavailability [299].

The ability to predict the incidence of many human activities with regard to a species and above all, to an ecosystem, is very limited. The difficulty in establishing a cause-and-effect relationship derives above all from the systematic lack of information on the "state of health" of the environment being studied or from the nature of its biological processes (which do not have linear characteristics and which are discontinuous through space and time). From here stems the importance of environmental monitoring plans and biomonitoring plans that, if properly applied, can supply an overall complete picture of the possible interventions that may be required.

Although lichens are important sources for control and environmental biomonitoring, it is necessary to take various precautions when using lichens as a quantitative measure of a single contaminant. Bioindicators of air pollution can most certainly supply information of a qualitative type; nonetheless, correlation studies using environmental data from the sites concerned and taken over a prolonged period of time (months–years), can supply information about semi-quantitative aspects.

Through IAP calculation, lichens allow us to evaluate air quality as far as regards the presence of different environmental contaminants. As already mentioned, lichens do not react specifically to a particular contaminant, but rather to the overall toxic effect of a mix of contaminants [54]. Levin and Pignata [230] proposed the use of Pollution Index (PI) for the evaluation of air quality. PI is determined using the equation cited by Levin and Pignata [230] and Pignata *et al.* [130] and Pignata [132]. This enables the evaluation of which of the biomonitored areas have the better air quality by measuring lichen reactions to atmospheric pollutants.

$$PI = \left[\left(\frac{\mathbf{Pha}}{\mathbf{Chla}} \right) + \left(\frac{\mathbf{S}_p}{\mathbf{S}_c} \right) \right] \left(\frac{\mathbf{HPCD}_p}{\mathbf{HPCD}_c} \right)$$

The Ph *a*/Chl *a* ratio can be changed to Chl *b*/Chl *a*:

$$PI = \left[\left(\frac{\mathbf{Chlb}}{\mathbf{Chla}} \right) + \left(\frac{\mathbf{S}_p}{\mathbf{S}_c} \right) \right] \left(\frac{\mathbf{HPCD}_p}{\mathbf{HPCD}_c} \right)$$

Chl b and **Chl a** express chlorophyll *a* and *b* concentrations in mg g^{-1} of dry weight. **S** is the sulphur content of lichens expressed as mg g^{-1} of dry weight, while **HPCD** expresses the concentration of hydroperoxy conjugated dienes in mmol g^{-1} of dry weight. The sub-index *p* indicates concentrations measured in samples transplanted to contaminated sites, while sub-index *c* indicates those measured in lichens transplanted to the control site.

Biomonitoring studies using lichens make it possible to verify, with our current state of knowledge, air quality and any improvements thereof. This is what happened in the case of the progressive improvement in air quality over the years (1989–1994) in several Italian cities (La Spezia) [67]. This improvement was due to the reduction in SO_2 emissions, which was partly linked to the increase in the use of methane gas for domestic heating and to the closure of a coal-fired power station. The city of Montecatini Terme (central Italy) has also improved its environmental situation: new lichen species have been found and the previous "lichen desert" situation has disappeared [66]. This phenomenon has been correlated with low SO_2 emission levels (approx 15–20 µg m^{-3} from 1993 to 1996) and NO_x, which passed from 150 µg m^{-3} in 1993 to 100 µg m^{-3} in 1996. A similar marked improvement caused by decrease in environmental SO_2 levels was also found in Paris in the 1980s (Luxembourg gardens) where lichen species from the previous century began to reappear [300, 301].

In general, lichen distribution in northern Italy seems mainly to be regulated by SO_2 pollution [68, 69, 302]. As far as regards central Italy, a study of *Parmelia caperata* made by Ref. [21] found a strong correlation ($r^2 = 0.93$; $p < 0.05$) between IAP values and the total heavy metal content (Cd, Cr, Cu, Hg, Ni, Pb, Zn).

Techniques for drawing up air quality maps using lichens, or the use of the transplant method, allow us to obtain information about a vast area in a short amount of time and at contained costs.

These methodological approaches, although they cannot be considered as replacements for standard atmospheric pollution monitoring carried out using control stations, are without a doubt valid environmental biomonitoring instruments in different cases:

1. as a preliminary evaluation, or rather as an estimate of the base impact in a set area, with the aim of preventing future human-derived impact;
2. to monitor an already-compromised environmental situation;
3. to control the quality of reclamation efforts already carried out.

Application of the system approach to the solving of problems regarding atmospheric pollution is doubtless valid and fundamentally requires an evaluation of the progress made in the areas of study considered, the identification of pollution sources and the cause/effect correlation of the same [298]. Of course, from that mentioned in the above points 1, 2 and 3, it can be seen that for this reason, necessary interventions must have three main objectives:

1. environmental prevention: with the aim of intervening at the impact source and thus in advance of the pollutant event;
2. environmental protection: to eliminate the effects of pollutant actions or to tend to minimise these effects;
3. environmental restoration: with the aim of removing damages caused by previous actions.

The necessity to increase our knowledge of bioindication studies using lichens remains a fundamental point in the development of research. It is possible to say that for a large majority of pollutants and their effects upon lichens, our knowledge is at an advanced stage in its development in terms of both the quantity and quality of information.

Nonetheless, it is possible to point out that in a significant part of bioindication studies of lichens, there is a tendency to study the environmental effects of situations that have already been compromised. This signifies a scarce propensity to carry out studies that fundamentally have an eye to aspects of environmental prevention.

References

[1] Hale M.E. (1969) How to Know the Lichens. Wm. C. Brown Company Publishers, Dubuque, Iowa, p. 226.

[2] Hale M.E. (1983) The Biology of Lichens. E. Arnold, London.

[3] Conti M.E., Cecchetti G. (2001) Biological monitoring: lichens as bioindicators of air pollution assessment – a review. Environmental Pollution, 114, 471–492, Reprinted (or higher parts taken) with a kind permission from Elsevier.

[4] Nylander W. (1866) Les lichens du Jardin du Luxembourg. Bulletin de La Société Botanique de France, 13, 364–372.

[5] Ferry B.W., Baddeley M.S., Hawksworth D.L. (1973) Air Pollution and Lichens. The Athlone Press, London, p. 389.

[6] Nimis P.L., Ciccarelli A., Lazzarin G., Barbagli R., Benedet A., Castello M., Gasparo D., Lausi D., Olivieri S., Tretiach M. (1989) I licheni come bioindicatori di inquinamento atmosferico nell'area di Schio-Thiene-Breganze (VI), in: Bolletino del Museo Civico di Storia Naturale di Verona, 16. CO.GE.V. s.r.l., Verona, Ecothema s.r.l., Trieste.

[7] Brodo I.M. (1961) Transplant experiments with coricolous lichens using a new technique. Ecology, 42, 838–841.

[8] Rao D.N., LeBlanc F. (1966) Effects of sulphur dioxide on the lichen alga, with special reference to chlorophyll. Bryologist, 69, 69–75.

[9] Schonbek H. (1968) Influence of air pollution (SO_2) on transplanted lichens. Naturwissenshaften, 55(9), 451–452.

[10] Hawksworth D.L. (1971) Lichens as litmus for air pollution: a historical review. International Journal of Environmental Studies, 1(4), 28–96.

[11] Gilbert O.L. (1973) Lichens and air pollution, in: Ahmadjian V., Hale M.E. (Eds.), The Lichens. Academic Press, New York, pp. 443–472.

[12] Méndez O.I., Fournier L.A. (1980) Los líquenes como indicadores de la contaminación atmosférica en el area metropolitana de San José, Costa Rica. Revista de Biologia Tropical, 28(1), 31–39.

[13] Lerond M. (1984) Utilisation des lichens pour la cartographie et le suivi de la pollution atmospherique. Bulletin D Ecologie, 15(1), 7–11.

[14] St. Clair L.L., Fields R.D. (1986) A comprehensive approach to biomonitoring of air quality using lichens. A field study. American Journal of Botany, 73(5), 610.

[15] St. Clair L.L., Fields R.D., Nakanishi M. (1986) Biomonitoring of air quality using lichens. A field study. American Journal of Botany, 73(5), 610.

[16] Galun M., Ronen R. (1988) Interaction of lichens and pollutants. CRC Handbook of Lichenology, 3, 55–72.

[17] Showman R.E. (1988) Mapping air quality with lichens – The North American experience, in: Nash T.H. III, Wirth W. (Eds.), Lichens, Bryophytes and Air Quality. Bibl. Lichenol. (Vol. 30). Cramer in der Gebruder Borntraeger Verlagsbuchhandlung., Berlin, pp. 67–90.

[18] Nimis P.L. (1990) Air quality indicators and indices: the use of plants as bioindicators of monitoring air pollution, in: Colombo A.G., Premazzi G. (Eds.), Proc. Workshop on Indicators and Indices JRC, Ispra, Italy, 93–126.

[19] Oksanen J., Laara E., Zobel K. (1991) Statistical analysis of bioindicator value of epiphytic lichens. Lichenologist, 23(2), 167–180.

[20] Seaward M.R.D. (1992) Large-scale air pollution monitoring using lichens. GeoJournal, 28(4), 403.

[21] Loppi S., Chiti F., Corsini A., Bernardi L. (1992) Preliminary data on the integrated use of lichens as indicators and monitors of atmospheric pollutants in central Italy. Giornale Botanico Italiano, 126, 360.

[22] Halonen P., Hyvärinen M., Kauppi M. (1993) Emission related and repeated monitoring of element concentrations in the epiphytic lichen *Hypogymnia physodes* in a coastal area, W Finland. Annales Botanici Fennici, 30, 251–261.

[23] Gries C. (1996) Lichens as indicators of air pollution, in: Nash T.H. III (Ed.), Lichen Biology. Cambridge University Press, Cambridge, pp. 240–254.

[24] Loppi S. (1996) Lichen as bioindicators of geothermal air pollution in central Italy. Bryologist, 99(1), 41–48.

[25] Seaward M.R.D. (1996) Lichens and the environment, in: Sutton B. (Ed.), A Century of Micology. Cambridge University Press, UK, pp. 293–320.

[26] Hamada N., Miyawaki H. (1998) Lichens as bioindicators of air pollution. Japanese Journal of Ecology, 48(1), 49–60.

[27] Asta J., Erhardt W., Ferretti M., Fornasier F., Kirschbaum U., Nimis P.L., Purvis O.W., Pirintsos S., Scheidegger C., van Haluwyn C., Wirth V. (2002) Mapping lichen diversity as an indicator of environmental quality, in: Nimis P.L., Scheidegger C., Wolseley P.A. (Eds.), Monitoring with Lichens – Monitoring Lichens. Kluwer Academic, Dordrecht, Boston, London, pp. 273–279.

[28] Nimis P.L., Scheidegger C., Wolseley P. (Eds.), (2002) Monitoring with Lichens-Monitoring Lichens. Kluwer Academic, Dordrecht, p. 408.

[29] Szczepaniak K., Biziuk M. (2003) Aspects of the biomonitoring studies using mosses and lichens as indicators of metal pollution. Environmental Research, 93(3), 221–230.

[30] Rossbach M., Lambrecht S. (2006) Lichens as biomonitors: global, regional and local aspects. Croatica Chemica Acta CCACAA, 79(1), 119–124.

[31] Ronen R., Canaani O., Garhy J., Cahen D., Malkin S., Galun M. (1984) The effect of air pollution and bisulphite treatment in the lichen *Ramalina*

duriaei studied by photoacoustics, in: Advances in Photosynthesis Research, Proceedings of the 6th Congress on Photosynthesis, 1–6 August 1983, Brussels.

[32] Calatayud A., Deltoro V.I., Abadia A., Abadia J., Barreno E. (1999) Effects on ascorbate feeding on chlorophyll fluorescence and xanthophyll cycle components in the lichen *Parmelia quercina* (Willd.) Vainio exposed to atmospheric pollutants. Physiologia Plantarum, 105(4), 679–684.

[33] Weissman L., Fraiberg M., Shine L., Garty J., Hochman A. (2006) Responses of antioxidants in the lichen *Ramalina lacera* may serve as an early-warning bioindicator system for the detection of air pollution stress. FEMS Microbiol Ecology, 58, 41–53.

[34] Kardish N., Ronen R., Bubrick P., Garty J. (1987) The influence of air pollution on the concentration of ATP and on chlorophyll degradation in the lichen *Ramalina duriaei* (De Not.) Bagl. New Phytologist, 106, 697–706.

[35] Garty J., Kardish N., Hagemeyer J., Ronen R. (1988) Correlations between the concentration of adenosine triphosphate, chlorophyll degradation and the amounts of airborne heavy metals and sulphur in a transplanted lichen. Archives of Environmental Contamination and Toxicology, 17(6), 601–611.

[36] Balaguer L., Manrique E. (1991) Interaction between sulphur dioxide and nitrate in some lichens. Environmental and Experimental Botany, 31(2), 223–227.

[37] Zaharopoulou A., Lanaras T., Arianoutsou M. (1993) Influence of dust from a limestone quarry on chlorophyll degradation of the lichen *Physcia adscendens* (Fr.) Oliv. Bulletin of Environmental Contamination and Toxicology, 50(6), 852–855.

[38] Epstein E., Sagee O., Cohen J.D., Garty J. (1986) Endogenous auxin and ethylene in the lichen *Ramalina duriaei*. Plant Physiology, 82, 1122–1125.

[39] Garty J., Karary Y., Harel J., Lurie S. (1993) Temporal and spatial fluctuations of ethylene production and concentrations of sulphur, sodium, chlorine and iron on/in the thallus cortex in the lichen *Ramalina duriaei* (De Not.) Bagl. Environmental and Experimental Botany, 33(4), 553–563.

[40] Conti M.E. (2002) Il monitoraggio biologico della qualità ambientale. SEAM, Roma, p. 180.

[41] Fields R.D., St. Clair L.L. (1984) The effects of SO_2 on photosynthesis and carbohydrate transfer in the two lichens: *Colema polycarpon* and *Parmelia chlorochroa*. American Journal of Botany, 71, 986–998.

[42] Rope S.K., Pearson L.C. (1990) Lichens as air pollution biomonitors in a semiarid environment in Idaho. The Bryologist, 93, 50–61.

[43] Silberstein L., Siegel B.Z., Siegel S.M., Mukhtar A., Galun M. (1996) Comparative studies on *Xanthoria parietina*, a pollution-resistant lichen, and *Ramalina duriaei*, a sensitive species. Evaluation of possible air pollution-protection mechanisms. Lichenologist, 28, 367–383.

[44] Takala K., Olkkonen H., Ikonen J., Jaaskelainen J., Puumalainen P. (1985) Total sulphur contents of epiphytic and terricolous lichens in Finland. Annales Botanici Fennici, 2, 91–100.

[45] Haffner E., Lomský B., Hynek V., Hallgren J.E., Batic F., Pfanz H. (2001) Air pollution and lichen physiology. Physiological responses of different lichens in a transplant experiment following an SO_2-gradient. Water Air and Soil Pollution, 131(1–4), 185–201.

[46] Carreras H.A., Gudiño G.L., Pignata M.L. (1998) Comparative biomonitoring of atmospheric quality in five zones of Córdoba city (Argentina) employing the transplanted lichen *Usnea sp.* Environmental Pollution, 103(2–3), 317–325.

[47] Ronen R., Galun M. (1984) Pigment extraction from lichens with dimethyl sulfoxide (DMSO) and estimation of chlorophyll degradation. Environmental and Experimental Botany, 24, 239–245.

[48] LeBlanc F., Rao D.N. (1975) Effects of air pollutants on lichens and bryophytes, in: Mudd B.J., Koziowski T.T. (Eds.), Responses of Plants to Air Pollution. Academic Press, pp. 237–271.

[49] Zambrano A., Nash T.H. III (2000) Lichen responses to short-term transplantation in Desierto de los Leones, Mexico City. Environmental Pollution, 107, 407–412.

[50] Rothe H., Bigdon M. (1994) Incidence of lichens in the area of the Hamburg airport. Gesundheitswesen, 56(10), 563–566.

[51] Richardson D.H.S. (1991) Lichens as biological indicators. Recent developments, in: Jeffrey D.W., Madden B. (Eds.), Bioindicators and Environmental Management. Academic Press, London, pp. 263–272.

[52] Seaward M.R.D. (1993) Lichens and sulphur dioxide air pollution: field studies. Environmental Reviews, 1, 73–91.

[53] LeBlanc F., De Sloover J. (1970) Relation between industrialization and the distribution and growth of epiphytic lichens and mosses in Montreal. Canadian Journal of Botany, 48, 1485–1496.

[54] Amman K., Herzig R., Liebendörfer L., Urech M. (1987) Multivariate correlation of deposition data of 8 different air pollutants to lichen data in a small town in Switzerland. Advances in Aerobiology, 87, 401–406.

[55] Herzig R., Urech M. (1991) Flechten als Bioindikatoren, integriertes biologisches Messystem der Luftverschmutzung für das Schweizer Mittelland. Bibliotheca Lichenologica, 43, 1–283.

[56] Lo Porto A., Macchiato M., Ragosta M. (1992) Bioindicazione della qualità dell'aria tramite licheni epifiti nella provincia di Potenza. Acqua & Aria, 1(92), 11–18.

[57] Gottardini E., Cristofolini F., Marchetti F. (1999) Biomonitoraggio della qualità dell'aria della città di Trento tramite licheni epifiti. Acqua & Aria, 4, 67–71.

[58] Pinho P., Augusto S., Branquinho C., Bio A., Pereira M.J., Soares A., Catarino F. (2004) Mapping lichen diversity as a first step for air quality assessment. Journal of Atmospheric Chemistry, 49(1–3), 377–389.

[59] Kommission Reinhaltung der Luft im VDI und DIN (1995) Messen von Immissionswirkungen. Measurement of Immission Effects, Dusseldorf.

[60] Zechmeister H.G., Hohenwallner D. (2006) A comparison of biomonitoring methods for the estimation of atmospheric pollutants in an industrial town in Austria. Environmental Monitoring and Assessment, 117, 245–259.

[61] Manuppella A., Carlomagno C. (1990) Air pollution and zonation of epiphytic lichens in the city of Isernia. Annali di igiene: medicina preventiva e di comunità, 2(5), 335–341.

[62] Nimis P.L. (1985) Urban lichen studies in Italy. 1st: the Town of Trieste. Geobotania, 5, 49–74.

[63] Nimis P.L. (1986) Urban lichen studies in Italy. 2nd: the Town of Udine. Geobotania, 7, 147–172.

[64] Loppi S., Corsini A., Chiti F., Bernardi L. (1992) Air quality bioindication by epiphytic lichens in central-northern Italy. Allionia, 31, 107–119.

[65] Monaci F., Bargagli R., Gasparo D. (1997) Air pollution monitoring by lichens in a small medieval town of central Italy. Acta Botanica Neerlandica, 46(4), 403–412.

[66] Loppi S., Giovannelli L., Pirintsos S.A., Putorti E., Corsini A. (1997) Lichen as bioindicators of recent changes in air quality (Montecatini Terme, Italy). Ecologia Mediterranea, 23(3–4), 53–56.

[67] Palmieri F., Neri R., Benco C., Serracca L. (1997) Lichens and moss as bioindicators and bioaccumulators in air pollution monitoring. Journal of Environmental Pathology, Toxicology and Oncology, 16(2–3), 175–190.

[68] Nimis P.L., Castello M., Perotti M. (1990) Lichens as biomonitors of sulphur dioxide pollutions in La Spezia (Northern Italy). Lichenologist, 22, 333–344.

[69] Nimis P.L., Lazzarin A., Lazzarin G., Gasparo D. (1991) Lichens as bioindicators of air pollution by SO_2 in the Veneto region (NE Italy). Studia Geobotanica, 11, 3–76.

[70] Piervittori R. (1998) Biomonitoring with lichens in the lower Susa Valley, Piedmont (Italy). Acta Horticulturae, 457, 319.

[71] Loppi S., Pacioni G., Olivieri N., Di Giacomo F. (1998) Accumulation of trace metals in the lichen Evernia prunastri transplanted at biomonitoring sites in central Italy. Bryologist, 101(3), 451–454.

[72] Grasso M.F., Clocchiatti R., Carrot F., Deschamps C., Vurro F. (1999) Lichens as bioindicators in volcanic areas: Mt Etna and Vulcano Island (Italy). Environmental Geology (Berlin), 37(3), 207–217.

[73] Brusoni M., Garavani M., Valcuvia Passadore M. (1997) Lichens and air pollution: preliminary studies in the Oltrepo Pavese (Pavia, Lombardy). Archivio Geobotanico, 3(1), 95–106.

[74] Bartoli A., Cardarelli E., Achilli M., Campanella L., Ravera S., Massari G. (1997) Quality assessment of the Maremma Laziale area using epiphytic lichens. Allionia (Turin), 35(0), 69–85.

[75] Cislaghi C., Nimis P.L. (1997) Lichens, air pollution and lung cancer. Nature, 387, 463–464.

[76] Nimis P.L., Tretiach M. (1995) The lichens of Italy – a phytoclimatical outline. Cryptogamic Botany, 5, 199–208.

[77] Nimis P.L. (1987) I macrolicheni d'Italia, chiavi analitiche per la determinazione. Atti del Museo Friulano di Storia Naturale, Udine, 8, 101–220.

[78] Wirth V. (1991) Zeigerwerte von Flechten. Scripta Geobot, 18, 215–237.

[79] Hawksworth D.L., Rose F. (1970) Qualitative scale for estimating sulphur dioxide air pollution in England and Wales using epiphytic lichens. Nature, 227(254), 145–148.

[80] Deruelle S. (1978) Les lichens et la pollution atmosphérique. Bulletin d' Ecologie, 9(2), 87–128.

[81] Van Haluwyn C., Lerond M. (1986) Les lichens et la qualité de l'air. Évolution méthodologique et limites. Ministère de l'environnement, SRETIE, p. 207.

[82] Lerond M., Van Haluwyn C., Cuny D. (1996) Lichens et bioindication: réalisations concrètes et exigences éthiques. Ecologie, 27(4), 277–283.

[83] Indice di Biodiversità Lichenica (2001) Manuale Agenzia Nazionale Protezione Ambiente (ANPA). Manuale e linee guida, Roma, p. 90.

[84] Trass H. (1973) Lichen sensitivity to the air pollution and index of poleotolerance (I.P.). Folia Cryptogamica Estonica, Tartu, 3, 19–22.

[85] Van Haluwyn C., Lerond M. (1988) Lichénosociologie et qualité de l'air: protocole opératoire et limites. Cryptogamie Bryologie Lichenologie, 9(4), 313–336.

[86] Khalil K., Asta J. (1998) Les lichens, bioindicateurs de pollution atmosphérique dans la région Lyonnaise. Ecologie, 29(3), 467–472.

[87] Insarov G.E., Semenov S.M., Insarova I.D. (1999) A system to monitor climate change with epilithic lichens. Environmental Monitoring and Assessment, 55(2), 279–298.

[88] Conti M.E., Tudino M., Stripeikis J., Cecchetti G. (2004) Heavy metal accumulation in the lichen *Evernia prunastri* transplanted at urban, rural and industrial sites in central Italy. Journal of Atmospheric Chemistry, 49, 83–94.

[89] Frati L, Brunialti G, Loppi S. (2005) Problems related to the transplants to monitor trace element deposition in repeated surveys: a case study from central Italy. Journal of Atmospheric Chemistry, 52, 221–30.

[90] James P.W. (1973) The effect of air pollutants other than hydrogen fluorides and sulphur dioxide on lichens, in: Ferry B.W., Baddeley M.S., Hawksworth D.L. (Eds.), Air Pollution and Lichens. The Athlone Press, London, pp. 143–176.

[91] Nieboer E., Richardson D.H.S., Tomassini F.D. (1978) Mineral uptake and release by lichens: an overview. The Bryologist, 81(2), 226–246.

[92] Bargagli R., Nimis P.L., Monaci F. (1997) Lichen biomonitoring of trace element deposition in urban, industrial and reference areas of Italy. Journal Trace Elements Medicine Biology, 11(3), 173–175.

[93] Burton M.A.S., LeSueur P., Puckett K.J. (1981) Copper, nickel and thallium uptake by the lichen *Cladina rangiferina*. Canadian Journal of Botany, 59, 91–100.

[94] Burton M.A.S. (1986) Biological monitoring of environmental contaminants (Plants). Marc-Report no. 32. Monitoring and Assessment Research Center, London.

[95] Brown D.H., Beckett R.P. (1984) Uptake and effect of cations on lichen metabolism. Lichenologist, 16, 173–188.

[96] Nash T.H. III, Wirth V. (Eds.), (1988) Lichens, bryophytes and air quality. Bibliotheca Lichenologica, 30. Cramer in der Gebruder Borntraeger Verlagsbuchhandlung., Berlin, p. 297.

[97] Puckett K.J. (1988) Bryophytes and lichens as monitors of metal deposition, in: Nash T.H. III, Wirth W. (Eds.), Lichens, Bryophytes and Air Quality. Bibliotheca Lichenologica, 30. Cramer in der Gebruder Borntraeger Verlagsbuchhandlung, Berlin, pp. 231–267.

[98] Richardson D.H.S. (1988) Understanding the pollution sensitivity of lichens. Botanical Journal of the Linnean Society, 96, 31–43.

[99] Nash T.H. III (1989) Metal tolerance in lichens, in: Shaw A.J. (Ed.), Heavy Metal Tolerance in Plants: Evolutionary Aspects. CRC Press, Boca Raton, FL, pp. 119–131.

[100] Brown D.H. (1991) Mineral cycling and lichens: the physiological basis. Lichenologist, 23, 293–307.

[101] Deruelle S. (1992) Lead accumulation in lichens. (Accumulation du plomb par les lichens). Bulletin – Societe Botanique de France, Actualites Botaniques, 139(1), 99–109.

[102] Richardson D.H.S. (1992) Pollution monitoring with lichens. Naturalists' Handbooks 19. Richmond Publishing, Slough, England.

[103] Richardson D.H.S. (1995) Metal uptake in lichens. Symbiosis, 18, 119–127.

[104] Sloof J.E. (1995) Lichens as quantitative biomonitors for atmospheric trace-element deposition, using transplants. Atmospheric Environment, 29, 11–20.

[105] Garty J. (1992) Lichens and heavy metals in the environment, in: Vernet J.P. (Ed.), Impact of Heavy Metals on the Environment (Vol. 2), Elsevier, Amsterdam, pp. 55–131.

[106] Garty J. (1993) Lichens as biomonitors for heavy metal pollution, in: Markert B. (Ed.), Plants as Biomonitors: Indicators for Heavy Metals in the Terrestrial Environment. VCH Publishers, Weinheim and New York, pp. 193–263.

[107] Garty J., Kloog N., Cohen Y., Wolfson R., Karnieli A. (1997) The effect of air pollution on the integrity of chlorophyll, spectral reflectance response, and on concentration of nickel, vanadium and sulphur in the lichen Ramalina duriaei (De Not.) Bagl. Environmental Research, 74(2), 174–187.

[108] Garty J., Cohen Y., Kloog N. (1998) Airborne elements, cell membranes, and chlorophyll in transplanted lichens. Journal of Environmental Quality, 27, 973–979.

[109] Bennett J.P., Wetmore C.M. (1999) Changes in element contents of selected lichens over 11 years in northern Minnesota, USA. Environmental and Experimental Botany, 41(1), 75–82.

[110] Freitas M.C., Reis M.A., Alves L.C., Wolterbeek H.Th. (1999) Distribution in Portugal of some pollutants in the lichen Parmelia sulcata. Environmental Pollution, 106(2), 229–235.

[111] Freitas M.C., Reis M.A., Marques A.P., Wolterbeek H.Th. (2001) Use of lichen transplants in atmospheric deposition studies. Journal of Radioanalytical and Nuclear Chemistry, 249(2), 307–315.

[112] Reis M.A., Alves L.C., Freitas M.C., Van Os B., de Goeij J., Wolterbeek H.Th. (2002) Calibration of lichen transplants considering faint memory effects. Environmental Pollution (Oxford, United Kingdom), 120(1), 87–95.

[113] Adamo P., Giordano S., Vingiani S., Castaldo Cobianchi R., Violante P., (2003) Trace element accumulation by moss and lichen exposed in bags in the city of Naples (Italy). Environmental Pollution, 122, 91–103.

[114] Marques A.P., Freitas M.C., Reis M.A., Wolterbeek H.Th., Verburg T. (2004) MCTTFA applied to differential biomonitoring in Sado estuary region. Journal of Radioanalytical and Nuclear Chemistry, 259(1), 35–40.

[115] Williamson B.J., Mikhailova I., Purvis O.W., Udachin V. (2004) SEM-EDX analysis in the source apportionment of particulate matter on *Hypogymnia physodes* lichen transplants around the Cu smelter and former mining town of Karabash, South Urals, Russia. Science of the Total Environment, 322, 139–154.

[116] Farinha M.M., Slejkovec Z., van Elteren J.T., Wolterbeek H.Th., Freitas M.C. (2004) Arsenic speciation in lichens and in coarse and fine airborne particulate matter by HPLC-UV-HG-AFS. Journal of Atmospheric Chemistry, 49(1–3), 343–353.

[117] Białonska D., Dayan F.E. (2005) Chemistry of the lichen *Hypogymnia physodes* transplanted to an industrial region. Journal of Chemical Ecology, 31(12), 2975–2991.

[118] Ayrault S., Clochiatti R., Carrot F., Daudin L., Bennett J.P. (2007) Factors to consider for trace element deposition biomonitoring surveys with lichen transplants. Science of the Total Environment, 372, 717–727.

[119] Baker D.A. (1983) Uptake of cations and their transport within the plants, in: Robb D.A., Pierpoint W.S. (Eds.), Metals and Micronutrients: Uptake and Utilization by Plants. Academic Press, London, pp. 3–19.

[120] Knops J.M.H., Nash T.H. III, Boucher V.L., Schlesinger W.L. (1991) Mineral cycling and epiphytic lichens: implications at the ecosystem level. Lichenologist, 23, 309–321.

[121] Nash T.H. III, Gries C. (1995) The use of lichens in atmospheric deposition studies with an emphasis on the Arctic. Science of the Total Environment, 160, 729–736.

[122] Antonelli M.L., Ercole P., Campanella L. (1998) Studies about the adsorption on lichen *Evernia prunastri* by enthalpimetric measurements. Talanta, 45, 1039–1047.

[123] Andersen A., Hovmand M.F., Johnsen I. (1978) Atmospheric heavy metal deposition in the Copenhagen area. Environmental Pollution, 17, 133–151.

[124] Herzig R. (1993) Plants as biomonitors – indicators for heavy metals in the terrestrial environment, in: Markert B. (Ed.), VCH Publishers, Weinheim and New York, pp. 286–328.

[125] Herzig R., Liebendörfer L., Urech M., Ammann K., Guecheva M., Landolt W. (1989) Passive biomonitoring with lichens as a part of an integrated biological measuring system for monitoring air pollution in Switzerland. International Journal of Environmental Analytical Chemistry, 35, 43–57.

[126] Sloof J.E., Wolterbeek H.Th. (1991) National trace element air pollution monitoring survey using epiphytic lichens. Lichenologist, 23(2), 139–166.

[127] Bari A., Minciardi M., Troiani F., Bonotto F., Paonessa F. (1998) Lichens and mosses in air quality monitoring: a biological model proposal. Govt. Reports Announcements & Index, Issue 16.

[128] Nimis P.L., Lazzarin G., Lazzarin A., Skert N. (2000) Biomonitoring of trace elements with lichens in Veneto (NE Italy). Science of the Total Environment, 255(1–3), 97–111.

[129] Panichev N., McCrindle R.I. (2004) The application of bio-indicators for the assessment of air pollution. Journal of Environmental Monitoring, 6(2), 121–123.

[130] Pignata M.L., González C.M., Wannaz E.D., Carreras H.A., Gudiño G.L. (2004) Biomonitoring of air quality employing in situ Ramalina celastri in Argentina. International Journal of Environment and Pollution, 22(4), 409–429.

[131] Vieira B.J., Freitas M.C., Rodrigues A.F., Pacheco A.M.G., Soares P.M., Correia N. (2004) Element-enrichment factors in lichens from Terceira, Santa Maria and Madeira Islands (Azores and Madeira archipelagoes). Journal of Atmospheric Chemistry, 49(1–3), 231–249.

[132] Pignata M.L. (2007) Distribution of atmospheric trace elements and assessment of air quality in Argentina employing the lichen, Ramalina celastri, as a passive biomonitor: detection of air pollution emission sources. International Journal of Environment and Health, 1(1), 29–46.

[133] Sheppard P.R., Speakman R.J., Ridenour G., Witten M.L. (2007) Using lichen chemistry to assess airborne tungsten and cobalt in Fallon, Nevada. Environmental Monitoring and Assessment, 130, 511–518.

[134] Bartoli A., Cardarelli E., Achilli M., Campanella L., Massari G. (1994) Biomonitoraggio dell'aria di Roma: accumulo di metalli pesanti in trapianti di licheni. Annals of Botany, LII(11), 239–266.

[135] Cardarelli E., Achilli M., Campanella C., Bartoli A. (1993) Monitoraggio dell'inquinamento da metalli pesanti mediante l'uso di licheni nella città di Roma. Inquinamento, 6, 56–63.

[136] Deruelle S. (1996) The reliability of lichens as biomonitors of lead pollution. Ecologie (Brunoy) 27(4), 285–290.

[137] Chettri M.K., Sawidis T. (1997) Impact of heavy metals on water loss from lichen talli. Ecotoxicology and Environmental Safety, 37(2), 103–111.

[138] Kłos A., Rajfur M., Wacławeka M., Wacławek W. (2006) Determination of the atmospheric precipitation pH value on the basis of the analysis of lichen cationoactive layer constitution. Electrochimica Acta, 51, 5053–5061.

[139] Kłos A., Rajfur M., Wacławek M., Wacławek W. (2005) Ion equilibrium in lichen surrounding. Bioelectrochemistry, 66(1–2), 95–103.

[140] Kral R., Kryzova L., Liska J. (1989) Background concentrations of lead and cadmium in the lichen Hypogymnia physodes at different altitudes. Science of the Total Environment, 84, 201–209.

[141] Jeran Z., Jacimov R., Batic F., Smodis B., Wolterbeek H.Th. (1996) Atmospheric heavy metal pollution in Slovenia derived from results for epiphytic lichens. Fresenius' Journal of Analytical Chemistry, 354(5–6), 681–687.

[142] Loppi S., Nelli L., Ancora S., Bargagli R. (1997) Passive monitoring of trace elements by means of tree leaves, epiphytic lichens and bark substrate. Environmental Monitoring and Assessment, 45(1), 81–88.

[143] Loppi S., Pirintsos S.A., De Dominicis V. (1999) Soil contribution to the elemental composition of epiphytic lichens (Tuscany, Central Italy). Environmental Monitoring and Assessment, 58, 121–131.

[144] Nieboer E., Richardson D.H.S., Lavoie P., Padovan D. (1979) The role of metal-ion binding in modifying the toxic effects of sulphur dioxide on the lichen *Umbilicaria muhlenbergii*. I. Potassium efflux studies. New Phytology, 82, 621–632.

[145] Beckett R.P., Brown D.H. (1984) The control of cadmium uptake in the lichen genus Peltigera. Journal of Experimental Botany, 35, 1071–1082.

[146] Riga-Karandinos A.N., Karandinos M.G. (1998) Assessment of air pollution from a lignite power plant in the plain of Megalopolis (Greece) using as biomonitors three species of lichens; impacts on some biochemical parameters of lichens. Science of the Total Environment, 215, 167–183.

[147] Poblet A., Andrade S., Scagliola M., Vodopivez C., Curtosi A., Pucci A., Marcovecchio J. (1997) The use of epiphytic Antarctic lichens (*Usnea aurantiacoatra* and *U. antarctica*) to determine deposition patterns of heavy metals in the Shetland Islands, Antarctica. Science of the Total Environment, 207(2–3), 187–194.

[148] Hyvarinen M., Crittenden P.D. (1996) Cation ratios in *Cladonia portentosa* as indices of precipitation acidity in the British Isles. New Phytologist, 132(3), 521–532.

[149] Walker T.R., Crittenden P.D., Young S.D., Prystina T. (2006) An assessment of pollution impacts due to the oil and gas industries in the Pechora basin, north-eastern European Russia. Ecological Indicators, 6, 369–387.

[150] Tarhanen S. (1998) Ultrastructural responses of the lichen *Bryoria fuscescens* to simulated acid rain and heavy metal deposition. Annals of Botany (London), 82(6), 735–746.

[151] Tarhanen S., Metssarinne S., Holopainen T., Oksanen J. (1999) Membrane permeability response of lichen *Bryoria fuscescens* to wet deposited heavy metals and acid rain. Environmental Pollution, 104(1), 121–129.

[152] Rossbach M., Jayasekera R., Kniewald G., Thang N.H. (1999) Large scale air monitoring: lichen vs. air particulate matter analysis. Science of the Total Environment, 232(1–2), 59–66.

[153] Kauppi M. (1976) Fruticose lichen transplant technique for air pollution experiments. Flora, 165, 407–414.

[154] Laaksovirta K., Olkkonen H. (1977) Epiphytic lichen vegetation and element contents of *Hypogymnia physodes* and pine needles examined as indicators of air pollution at Kokkola, W. Finland. Annales botanici Fennici, 14, 112–130.

[155] Palomaki V., Tynnyrinen S., Holopainen T. (1992) Lichen transplantation in monitoring fluoride and sulphur deposition in the surroundings of a fertilizer plant and a strip mine an Siilinjarvi. Annales Botanici Fennici, 29(1), 25–34.

[156] Pilegaard K. (1978) Airborne metals and SO_2 monitored by epiphytic lichens in an industrial area. Environmental Pollution, 17, 81–92.

[157] Pilegaard K. (1979) Heavy metals in bulk precipitation and transplanted Hymogimnia physodes and Dicranoweisia cirrata in the vicinity of a Danish steel works. Water Air Soil Pollution, 11, 77–91.

[158] Pilegaard K., Rasmussen L., Gydesen H. (1979) Atmospheric background deposition of heavy metals in Denmark monitored by epiphytic cryptogams. Journal of Applied Ecology, 16, 843–853.

[159] Addison P.A., Puckett K.J. (1980) Deposition of atmospheric pollutants as measured by lichen element content in the Athabasca oil sands area. Canadian Journal of Botany, 58, 2323–2334.

[160] Pakarinen P., Kaistila M., Hasanen E. (1983) Regional concentration levels of vanadium, aluminium and bromine in mosses and lichens. Chemosphere, 12, 1477–1485.

[161] De Bruin M., Hackenitz E. (1986) Trace elements concentrations in epiphytic lichens and bark substrate. Environmental Pollution (series B), 11, 153–160.

[162] Nieboer E., Ahmed H.M., Puckett K.J., Richardson D.H.S. (1972) Heavy metal content of lichens in relation to distance from a nickel smelter in Sudbury, Ontario. Lichenologist, 5, 291–304.

[163] Olmez I., Gulovali M.C., Gordon G.E. (1985) Trace element concentrations in lichens near a coal-fired power plant. Atmospheric Environment – Part A General Topics, 19(10), 1663–1669.

[164] Garty J. (1987) Metal amounts in the lichen *Ramalina duriaei* (De Not.) Bagl. transplanted at biomonitoring sites around a new coal-fired power station after 1 year of operation. Environmental Research, 43, 104–116.

[165] Freitas M.C. (1994) Heavy metals in *Parmelia sulcata* collected in the neighborhood of a coal-fired power station. Biological Trace Element Research, 43–45, 207–212.

[166] Gonzalez C.M., Pignata M.L. (1997) Chemical response of the lichen *Punctelia subrudecta* (Nyl.) Krog transplanted close to a power station in an urban-industrial environment. Environmental Pollution, 97(3), 195–203.

[167] Cuny D., Davranche L., Thomas P., Kempa M., Van Haluwyn C. (2004) Spatial and temporal variations of trace element contents in *Xanthoria Parietina* thalli collected in a highly industrialized area in Northern France as an element for a future epidemiological study. Journal of Atmospheric Chemistry, 49(1–3), 391–401.

[168] Branquinho C., Catarino F., Brown D.H., Pereira M.J., Soares A. (1999) Improving the use of lichens as biomonitors of atmospheric metal pollution. Science of the Total Environment, 232(1–2), 67–77.

[169] Chettri M.K., Sawidis T., Karataglis S. (1997) Lichens as a tool for biogeochemical prospecting. Ecotoxicology and Environmental Safety, 38(3), 322–335.

[170] Monnet F., Bordas F., Deluchat V., Chatenet P., Botineau M., Baudu M. (2005) Use of the aquatic lichen *Dermatocarpon luridum* as bioindicator of copper pollution: accumulation and cellular distribution tests. Environmental Pollution, 138, 455–461.

[171] Cabral J.P. (2003) Copper toxicity to five *Parmelia* lichens *in vitro*. Environmental and Experimental Botany, 49, 237–250.

[172] Lupsina V., Horvat M., Jeran Z., Stegnar P. (1992) Investigation of mercury speciation in lichens. Analyst, 117(3), 673–675.

[173] Balarama Krishna M.V., Karunasagar D., Arunachalam J. (2003) Study of mercury pollution near a thermometer factory using lichens and mosses. Environmental Pollution, 124, 357–360.

[174] Sánchez Uría J.E., Sanz-Medel A. (1998) Inorganic and methylmercury speciation in environmental samples. Talanta, 47, 509–524.

[175] Balarama Krishna M.V., Manjusha Ranjit, Karunasagar D., Arunachalam J. (2005) A rapid ultrasound-assisted thiourea extraction method for the determination of inorganic and methyl mercury in biological and environmental samples by CVAAS. Talanta, 67, 70–80.

[176] Balarama Krishna M.V., Karunasagar D., Arunachalam J. (2004) Sorption characteristics of inorganic, methyl and elemental mercury on lichens and mosses: implication in biogeochemical cycling of mercury. Journal of Atmospheric Chemistry, 49, 317–328.

[177] Bargagli R., Barghigiani C. (1991) Lichen biomonitoring of mercury emission and deposition in mining, geothermal and volcanic areas of Italy. Environmental Monitoring Assessment, 16, 265–275.

[178] Koranda J.J. (1980) Studies of boron deposition near geothermal power plants. U.S. Department of Energy, Interim Report, UCID 18606, Berkeley.

[179] Matthews K.M. (1981) The use of lichens in a study of geothermal radon emissions in New Zealand. Environmental Pollution, 24, 105–116.

[180] Connor J.J. (1979) Geochemistry of ohia and soil lichen, Puhimau thermal area, Hawaii. Science of the Total Environment, 12(3), 241–250.

[181] Loppi S., Cenni E., Bussotti F., Ferretti M. (1998) Biomonitoring of geothermal air pollution by epiphytic lichens and forest trees. Chemosphere, 36(4–5), 1079–1082.

[182] Loppi S., Bargagli R. (1996) Lichen biomonitoring of trace elements in a geothermal area (central Italy). Water Air and Soil Pollution, 88(1–2), 177–187.

[183] Beauchamp R.O., Bus J.S., Popp J.A., Boreiko C.J., Andjelkovich D.A. (1984) A critical review of the literature on hydrogen sulphide toxicity. Critical Review in Toxicology, 13, 25–97.

[184] Garty J., Kloog N., Cohen Y. (1998) Integrity of lichen cell membranes in relation to concentration of airborne elements. Archives of Environmental Contamination and Toxicology, 34(2), 136–144.

[185] Pearson L.C., Henriksson E. (1981) Air pollution damage to cell membranes in lichens. II. Laboratory experiments. Bryologist, 84, 515–520.

[186] Hart R., Webb P.G., Biggs R.H., Portier K.M. (1988) The use of lichen fumigation studies to evaluate the effects of new emission sources on class I areas. JAPCA, 38, 144–147.

[187] Alebic-Juretic A., Arko-Pijevac M. (2005) Lichens as indicators of air pollution in the city of Rijeka, Croatia. Fresenius Environmental Bulletin, 14(1), 40–43.

[188] Marques A.P., Freitas M.C., Wolterbeek H.Th., Steinebach O.M., Verburg T., De Goeij J.J.M. (2005) Cell-membrane damage and element leaching in transplanted *Parmelia sulcata* lichen related to ambient SO_2, temperature, and precipitation. Environmental Science and Technology, 39(8), 2624–2630.

[189] Godinho R.M., Freitas M.C., Wolterbeek H.Th. (2004) Assessment of lichen vitality during a transplantation experiment to a polluted site. Journal of Atmospheric Chemistry, 49, 355–361.

[190] Freitas M.C., Pacheco A.M.G. (2004) Bioaccumulation of cobalt in *Parmelia sulcata*. Journal of Atmospheric Chemistry, 49, 67–82.

[191] Carreras H.A., Wannaz E.D., Perez C.A., Pignata M.L. (2005) The role of urban air pollutants on the performance of heavy metal accumulation in *Usnea amblyoclada*. Environmental Research, 97(1), 50–57.

[192] Davies F., Notcutt G. (1996) Biomonitoring of atmospheric mercury in the vicinity of Kilauea, Hawaii. Water Air and Soil Pollution, 86(1–4), 275–281.

[193] Wolterbeek H.Th., Bode P. (1995) Strategies in sampling and sample handling in the context of large-scale plant biomonitoring surveys of trace element air pollution. Science of the Total Environment, 176(1–3), 33–43.

[194] Quevauviller P., Herzig R., Muntau H. (1996) Certified reference material of lichen (CRM 482) for the quality control of trace element biomonitoring. Science of the Total Environment, 187, 143–152.

[195] Smodis B., Parr R.M. (1999) Biomonitoring of air pollution as exemplified by recent IAEA programs. Biological Trace Element Research, 71–72, 257–266.

[196] Acar O., Ozvatan S., Ilim M. (2005) Determination of cadmium, copper, iron, manganese, lead and zinc in lichens and botanic samples by electrothermal and flame atomic absorption spectrometry. Ankara Nuclear Research and Training Center, Ankara, Turk. Turkish Journal of Chemistry, 29(4), 335–344.

[197] Rossbach M., Zeiller E. (2003) Assessment of element-specific homogeneity in reference materials using microanalytical techniques. Department of Nuclear Science and Applications, International Atomic Energy Agency, Vienna, Austria. Analytical and Bioanalytical Chemistry, 377(2), 334–339.

[198] Sha Y., Zhang P., Wang X., Liu J., Huang Y., Li G. (2002) Analysis of candidate micro-reference materials of lichen and algae by SRXRF and PIXE. Nuclear Instruments & Methods in Physics Research, Section B: Beam Interactions with Materials and Atoms, 189, 107–112.

[199] Dybczynski R., Danko B., Polkowska-Motrenko H. (2000) NAA study on homogeneity of reference materials and their suitability for microanalytical techniques. Journal of Radioanalytical and Nuclear Chemistry, 245(1), 97–104.

[200] Marques A.P., Freitas M.C., Wolterbeek H.Th., Verburg T.G., De Goeij J.J.M. (2007) Grain – size effects on PIXE and INAA analysis of IAEA – 336 lichen reference material. Nuclear Instruments & Methods in Physics Research, Section B: Beam Interactions with Materials and Atoms, 255(2), 380–394.

[201] Jackson L.L., Ford J., Schwartzman D. (1993) Collection and chemical analysis of lichen for biomonitoring. Government Reports Announcements and Index, 09/93.

[202] Wadleigh M.A., Blake D.M. (1999) Tracing sources of atmospheric sulphur using epiphytic lichens. Environmental Pollution, 106, 265–271.

[203] Conti M.E., Tudino M.B., Muse J.O., Cecchetti G.F. (2002) Biomonitoring of heavy metals and their species in the marine environment: the contribution of atomic absorption spectroscopy and inductively coupled plasma spectroscopy. Research Trends in Applied Spectroscopy, 4, 295–324. Reprinted (or higher parts taken) with a kind permission from Research Trends.

[204] Bocca B., Conti M.E., Pino A., Mattei D., Forte G., Alimonti A. (2007) Simple, fast and clean digestion procedures for determination of chemical elements in biological and environmental matrices by SF-ICP-MS. International Journal Environmental Analytical Chemistry, 87(15), 1111–1123.

[205] Pino A., Alimonti A., Botrè F., Minoia C., Bocca B., Conti M.E. (2007) Determination of twenty-five elements in lichens by sector field inductively coupled plasma mass spectrometry and microwave-assisted acid digestion. Rapid Communications in Mass Spectrometry, 21(12), 1900–1906.

[206] Balarama Krishna M.V., Arunachalam J. (2004) Ultrasound-assisted extraction procedure for the fast estimation of major, minor and trace elements in lichen and mussel samples by ICP-MS and ICP-AES. Analytica Chimica Acta, 522, 179–187.

[207] Paul A., Hauck M. (2006) Effects of manganese on chlorophyll fluorescence in epiphytic cyano- and chlorolichens. Flora: Morphology, Distribution, Functional Ecology of Plants, 201(6), 451–460.

[208] Salemaa M., Derome J., Helmisaari H.-S., Nieminen T., Vanha-Majamaa I. (2004) Element accumulation in boreal bryophytes, lichens and vascular plants exposed to heavy metal and sulfur deposition in Finland. Science of the Total Environment, 324, 141–160.

[209] Grass F., Bichler M., Dorner J., Holzner H., Ritschel A., Ramadan A., Westphal G.P., Gwozdz R. (1994) Application of short-lived radionuclides

in neutron activation analysis of biological and environmental samples. Biological Trace Element Research, 43–45, 33–46.

[210] Caniglia G., Calliari I., Celin L., Tollardo A.M. (1994) Metal determination by EDXFR in lichens. A contribution to pollutants monitoring. Biological Trace Element Research, 43–45, 213–221.

[211] Richardson D.H.S., Shore M., Richardson R.M. (1995) The use of X-ray fluorescence spectrometry for the analysis of plants, especially lichens, employed in biological monitoring. Science of the Total Environment, 176, 1–3, 97–105.

[212] Ranta P. (2001) Changes in urban lichen diversity after a fall in sulphur dioxide levels in the city of Tampere, SW Finland. Annales Botanici Fennici, 38(4), 295–304.

[213] Hauck M., Hesse V., Runge M. (2002) The significance of stemflow chemistry for epiphytic lichen diversity in a dieback-affected spruce forest on Mt Brocken, northern Germany. Lichenologist, 34(5), 415–427.

[214] Vingiani S., Adamo P., Giordano S. (2004) Sulphur, nitrogen and carbon content of *Sphagnum capillifolium* and *Pseudevernia furfuracea* exposed in bags in the Naples urban area. Environmental Pollution, 129, 145–158.

[215] Henriksson E., Pearson L.C. (1981) Nitrogen fixation rate and chlorophyll content of the lichen *Peltigera canina* exposed to sulphur dioxide. American Journal of Botany, 68, 680–684.

[216] Fields R.F. (1988) Physiological responses of lichens to air pollutant fumigations, in: Nash T.H. III, Wirth V. (Eds.), Lichens, Bryophytes and Air Quality. Bibliotheca Lichenologica, 30. Cramer in der Gebruder Borntraeger Verlagsbuchhandlung, Berlin, pp. 175–200.

[217] Gries C., Sanz M.J., Nash T.H. III (1995) The effect of SO_2 fumigation on CO_2 gas exchange, chlorophyll fluorescence and chlorophyll degradation in different lichen species from western North America. Cryptogamic Botany, 5(3), 239–246.

[218] Scott M.G., Hutchinson T.C. (1987) Effects of a simulated acid rain episode on photosynthesis and recovery in the caribou-forage lichens, *Cladina stellaris* (Opiz.) Brodo and *Cladina rangiferina* (l) Wigg. New Phytologist, 107, 567–575.

[219] Holopainen T., Kauppi M. (1989) A comparison of light fluorescence and electron microscopic observations in assessing the SO_2 injury of lichens under different moisture conditions. Lichenologist, 21, 119–134.

[220] Sanz M.J., Gries C., Nash T.H. III (1992) Dose-response relationships for SO_2 fumigations in the *Evernia prunastri* (L.) Ach. and *Ramalina fraxinea* (L.) Ach. New Phytology, 122, 313–319.

[221] Kong F., Hu W., Sang W., Wang L. (2002) Effects of sulphur dioxide on the relationship between symbionts in lichen. Chinese Journal of Applied Ecology, 13(2), 151–155.

[222] Baddeley M.S., Ferry B.W., Finegan E.J. (1972) The effect of sulphur dioxide on lichen respiration. Lichenologist, 5, 284–291.

[223] Richardson D.H.S, Puckett K.J. (1973) Sulphur dioxide and photosynthesis in lichens, in: Ferry B.W., Baddeley M.S., Hawksworth D.L. (Eds.), Air Pollution and Lichens. The Athlone Press, London, pp. 283–298.

[224] Lechowicz M.J. (1982) The effect of simulated acid precipitation on photosynthesis in the caribou lichen *Cladina stellaris* (Opiz.) Brodo. Water Air Soil Pollution, 18, 421–430.

[225] Deltoro V.I., Gimeno C., Calatayud A., Barreno E. (1999) Effects of SO_2 fumigations on photosynthetic CO_2 gas exchange, chlorophyll a fluorescence emission and antioxidant enzymes in the lichens *Evernia prunastri* and *Ramalina farinacea*. Physiologia Plantarum, 105(4), 648–654.

[226] Calatayud A., Sanz M.J., Calvo E., Barreno E., del Valle-Tascon (1996) Chlorophyll a fluorescence and chlorophyll content in *Parmelia quercina* thalli from a polluted region of northern Castellon (Spain). Lichenologist, 28, 49–65.

[227] Backor M., Pualíková K., Geralská A., Davidson R. (2003) Monitoring of air pollution in Kosice (Eastern Slovakia) using lichens. Polish Journal of Environmental Studies, 12(2), 141–150.

[228] Boonpragob K., Nash T.H. III (1991) Physiological responses of the lichen Ramalina menziesii Tayl. to the Los Angeles urban environment. Environmental and Experimental Botany, 31(2), 229–238.

[229] González C.M., Pignata M.L. (1994) The influence of air pollution on soluble proteins, chlorophyll degradation, MDA, sulphur and heavy metals in a transplanted lichen. Chemistry and Ecology, 9, 105–113.

[230] Levin A.G., Pignata M.L. (1995) *Ramalina ecklonii* (Spreng) Mey. and Flot. as bioindicator of atmospheric pollution in Argentina. Canadian Journal of Botany, 73(8), 1196–1202.

[231] González C.M., Casanovas S.S., Pignata M.L. (1996) Biomonitoring of air pollutants from traffic and industries employing *Ramalina ecklonii* (Spreng.) Mey. and Flot. in Córdoba, Argentina. Environmental Pollution, 91(3), 269–277.

[232] González C.M., Orellana L.C., Casanovas S.S., Pignata M.L. (1998) Environmental conditions and chemical response of a transplanted lichen to an urban area. Journal of Environmental Management, 53(1), 73–81.

[233] Yun M., Wadleigh M.A., Pye A. (2004) Direct measurement of sulphur isotopic composition in lichens by continuous flow-isotope ratio mass spectrometry. Chemical Geology, 204, 369–376.

[234] Insarova I.D., Insarov G.E., Semenov S.M., Braakenhielm S., Hultengren S. (1993) Lichen sensitivity and air pollution – a review of literature data. Govt Reports Announcements & Index, Issue 17, 1993.

[235] Hawksworth D.L., Rose F. (1976) Lichens as pollution monitors. Institut of Biology. Studies in biology, 66. E. Arnold, London.

[236] Garty J., Kauppi M., Kauppi A. (1995) Differential responses of certain lichen species to sulphur-containing solutions under acidic conditions as expressed by the production of stress ethylene. Environmental Research, 69(2), 132–143.

[237] Case J.W., Krouse H.R. (1980) Variations in sulphur content and stable sulphur isotope composition of vegetation near a SO$_2$ source at Fox Creek, Alberta, Canada. Oecologia, 44, 248–257.

[238] Krouse H.R., Case J.W. (1981) Sulphur isotope ratios in water, air and vegetation near Teepee Creek gas plant, Alberta. Water Air and Soil Pollution, 15, 11–28.

[239] Takala K., Olkkonen H., Krouse H.R. (1991) Sulphur isotope composition of epiphytic and terricolous lichens and pine bark in Finland. Environmental Pollution, 69, 337–348.

[240] Wadleigh M.A. (2003) Lichens and atmospheric sulphur: what stable isotopes reveal. Environmental Pollution, 126, 345–351.

[241] Wiseman R.D., Wadleigh M.A. (2002) Lichen response to changes in atmospheric sulphur: isotopic evidence. Environmental Pollution, 116, 235–241.

[242] Batts J.E., Calder L.J., Batts B.D. (2004) Utilizing stable isotope abundances of lichens to monitor environmental change. Chemical Geology, 204, 345–368.

[243] Satterwhite M.B., Ponder Henley J., Carney J.M. (1985) Effects of lichens on the reflectance spectra of granitic rock surfaces. Remote Sensing of Environment, 18, 105–112.

[244] Bradford M.M. (1976) A rapid and sensitive method for the quantitation of microgram quantities of proteins utilizing the principle of protein-dye binding. Analytical Biochemistry, 72, 248–254.

[245] Menzel D.B. (1976) The role of free radicals in the toxicity of air pollutants (nitrogen oxides and ozone), in: Pryor W.A. (Ed.), Free Radicals in Biology, 2. Academic Press, New York, pp. 181–203.

[246] Bychek-Guschina I.A., Kotlova E.R., Heipieper H. (1999) Effects of sulphur dioxide on lichen lipids and fatty acids. Biochemistry, 64(1), 61–65.

[247] Garty J., Weissman L., Levin T., Garty-Spitz R., Lehr H. (2004) Impact of UV-B, heat and chemicals on ethylene-production of lichens. Journal of Atmospheric Chemistry, 49(1–3), 251–266.

[248] De Bakker A.J. (1989) Effects of ammonia emission on epiphytic lichen vegetation. Acta Botanica Neerlandica, 38, 337–342.

[249] Van Dobben H.F., Ter Braak C.J.F. (1998) Effects of atmospheric NH$_3$ on epiphytic lichens in the Netherlands: the pitfalls of biological monitoring. Atmospheric Environment, 32(3), 551–557.

[250] Hyvarinen M., Crittenden P.D. (1998) Relationships between atmospheric nitrogen inputs and the vertical nitrogen and phosphorus concentration gradients in the lichen *Cladonia portentosa*. New Phytologist, 140(3), 519–530.

[251] Hyvarinen M., Crittenden P.D. (1998) Growth of the cushion-forming lichen, *Cladonia Portentosa*, at nitrogen-polluted and unpolluted heathland site. Environmental and Experimental Botany, 40(1), 67–76.

[252] Sochting U. (1995) Lichens as monitors of nitrogen deposition. Cryptogamic Botany, 5(3), 264–269.

[253] Bruteig I.E. (1994) Distribution, ecology and biomonitoring studies of epiphytic lichens on conifers. Gunneria, 68, p. 24.

[254] Gauslaa Y. (1985) The ecology of *Lobarion pulmonariae* and *Parmelion caperatae* in *Quercus* dominated forests in south-west Norway. Lichenologist, 17, 117–140.

[255] Gilbert O.L. (1986) Field evidence for an acid rain effect on lichens. Environmental Pollution (series A), 40, 227–231.

[256] Russow R., Veste M., Bohme F. (2005) A natural 15N approach to determine the biological fixation of atmospheric nitrogen by biological soil crusts of the Negev Desert. Rapid Communications in Mass Spectrometry, 19(23), 3451–3456.

[257] Gombert S., Asta J., Seaward M.R.D. (2006) Lichens and tobacco plants as complementary biomonitors of air pollution in the Grenoble area (Isère, southeast France). Ecological Indicators, 6, 429–443.

[258] Mehelman M.A., Borek C. (1987) Toxicity and biochemical mechanisms of ozone. Environmental Research, 42, 36–53.

[259] Egger R., Schlee D., Türk R. (1994) Changes of physiological and biochemical parameters in the lichen *Hypogymnia physodes* (L.) Nyl. due to the action of air pollutants – a field study. Phyton, 35, 229–242.

[260] Heath R.L., Packer L. (1968) Photoperoxidation in isolated chloroplast. I. Kinetics and stoichiometry of fatty acids peroxidation. Archives of Biochemistry and Biophysics, 125, 189–198.

[261] Kosugi H., Jojima T., Kikugawa K. (1989) Thiobarbituric acid-reactive substances from peroxidized lipids. Lipids, 24, 873–881.

[262] Scheidegger C., Schroeter B. (1995) Effects of ozone fumigation on epiphytic macrolichens: ultrastructure, CO_2 gas exchange and chlorophyll fluorescence. Environmental Pollution, 88, 345–354.

[263] Karlsson G.P., Sellden G., Pleijel H. (1995) Clover as an indicator plant for phytotoxic ozone concentrations: visible injury in relation to species, leaf age and exposure dynamics. New Phytologist, 129(2), 355.

[264] Benton J., Fuhrer J., Gimeno B.S., Skaerby L., Sanders G. (1995) Results from the UN/ECE ICP-Crops indicate the extent of exceedance of the critical levels of ozone in Europe. Water Air and Soil Pollution, 85(3), 1473–1478.

[265] Ross L.J., Nash T.H. III (1983) Effects of ozone on gross photosynthesis of lichens. Environmental and Experimental Botany, 23, 71–77.

[266] Asta J., Garrec J.P. (1980) Etude de l'accumulation du fluor dans les lichens d'une vallee Alpine polluee. Environmental Pollution (series A), 21, 267–286.

[267] Perkins D.F. (1992) Relationship between fluoride contents and loss of lichens near an aluminium works. Water Air Soil Pollution, 64(3–4), 503–510.

[268] Gombert S., Asta J. (1997) Monitoring the chlorine pollution of a refuse incinerator using lichens and sphagnum mosses. Ecologie (Brunoy), 28(4), 365–372.

[269] Schumacher M., Domingo J.L., Llobet J.M., Müller L., Jager J. (1997) Levels of PCDD/F in grasses and weeds collected near a municipal waste incinerator (1996–1997). Science of the Total Environment, 201, 53–62.

[270] Schumacher M., Domingo J.L., Llobet J.M., Sunderhauf W., Müller L. (1998) Temporal variation of PCDD/F concentrations in vegetation samples collected in the vicinity of a municipal waste incinerator (1996–1997). Science of the Total Environment, 218(2–3), 175–183.

[271] Augusto S., Pinho P., Branquinho C., Pereira M.J., Soares A., Catarino F. (2004) Atmospheric dioxin and furan deposition in relation to land-use and other pollutants: a survey with lichens. Journal of Atmospheric Chemistry, 49(1–3), 53–65.

[272] Oxynos S.K., Schmitzer K., Marth J., Kettrup P. (1997) PCDD/F and other chlorinated hydrocarbons in matrices of the Federal Environmental Specimen Bank. Chemosphere, 34(9–10), 2153–2158.

[273] Domeño C., Blasco M., Sánchez C., Nerín C. (2006) A fast extraction technique for extracting polycyclic aromatic hydrocarbons (PAHs) from lichens samples used as biomonitors of air pollution: Dynamic sonication versus other methods. Analytica Chimica Acta, 569, 103–112.

[274] Blasco M., Domeño C., Nerín C. (2006) Use of lichens as pollution biomonitors in remote areas: comparison of PAHs extracted from lichens and atmospheric particles sampled in and around the somport tunnel (pyrenees). Environmental Science and Technology, 40(20), 6384–6391.

[275] Favero-Longo S.E., Turci F., Tomatis M., Castelli D., Bonfante P., Hochella M.F., Piervittori R., Fubini B. (2005) Chrysotile asbestos is progressively converted into a non-fibrous amorphous material by the chelating action of lichen metabolites. Journal of Environmental Monitoring, 7(8), 764–766.

[276] Notter M. (1988) Radionuclides in the environment around Swedish nuclear power stations, 1983. Govt Reports Announcements & Index, 11, 1988.

[277] Barci G., Dalmasso J., Ardisson G. (1988) Chernobyl fallout measurements in some Mediterranean biotas. Science of the Total Environment, 70, 373–387.

[278] Seaward M.R.D., Heslop J.A., Green D., Bylinska E.A. (1988) Recent levels of radionuclides in lichens from southwest Poland with particular reference to cesium-134 and cesium-137. Journal of Environmental Radioactivity, 7(2), 123–130.

[279] Mihok S., Schwartz B., Wiewel A.M. (1989) Bioconcentration of fallout 137 Cs by fungi and red-backed voles (*Clethrionomys gapperi*). Health Physics, 57(6), 959–966.

[280] Livens F.R., Horrill A.D., Singleton D.L. (1991) Distribution of radiocesium in the soil-plant systems of upland areas of Europe. Health Physics, 60(4), 539–545.

[281] Sloof J.E., Wolterbeek H.Th. (1992) Lichens as biomonitors for radiocesium following the Chernobyl accident. Journal of Environmental Radioactivity, 16(3), 229–242.

[282] Hofmann W., Attarpour N., Lettner H., Turk R. (1993) 137 Cs concentrations in lichens before and after the Chernobyl accident. Health Physics, 64(1), 70–73.

[283] Triulzi C., Marzano F.N., Vaghi M. (1996) Important alpha, beta and gamma-emitting radionuclides in lichens and mosses collected in different world areas. Annali di Chimica, 86(11–12), 699–704.

[284] Sawidis T., Heinrich G., Chettri M.K. (1997) Cesium-137 monitoring using lichens from Macedonia, northern Greece. Canadian Journal of Botany, 75(12), 2216–2223.

[285] Kirchner G., Daillant O. (2002) The potential of lichens as long-term biomonitors of natural and artificial radionuclides. Environmental Pollution, 120, 145–150.

[286] Bretten S., Gaare E., Skogland T., Steinnes E. (1992) Investigations of radiocesium in the natural terrestrial environment in Norway following the Chernobyl accident. Analyst, 117(3), 501–503.

[287] Jones B.E., Eriksson O., Nordkvist M. (1989) Radiocesium uptake in reindeer on natural pasture. Science of the Total Environment, 85, 207–212.

[288] Rissanen K., Rahola T. (1989) Cs-137 concentration in reindeer and its fodder plants. Science of the Total Environment, 85, 199–206.

[289] Chant L.A., Andrews H.R., Cornett R.J., Koslowsky V., Milton J.C., Van den Berg G.J., Verburg T.G., Wolterbeek H.Th. (1996) 129I and 36Cl concentrations in lichens collected in 1990 from three regions around Chernobyl. Applied Radiation and Isotopes, 47(9–10), 933–937.

[290] Topcuoglu S., Van Dawen A.M., Gungor N. (1995) The natural depuration rate of 137 Cs radionuclides in a luchen and moss species. Journal of Environmental Radioactivity, 29(2), 157–162.

[291] Thomas R.S., Ibrahim S.A. (1995) Plutonium concentrations in lichen of Rocky Flats environs. Health Physics, 68(3), 311–319.

[292] Kwapulinski J., Seaward M.R., Bylinska E.A. (1985) Uptake of 226 radium and 228 radium by the lichen genus *Umbilicaria*. Science of the Total Environment, 41, 135–141.

[293] Golubev A.V., Golubeva V.N., Krylov N.G., Kuznetsova V.F., Mavrin S.V., Aleinikov A.Yu., Hoppes W.G., Surano K.A. (2005) On monitoring anthropogenic airborne uranium concentrations and 235U/238U isotopic ratio by lichen e bio-indicator technique. Journal of Environmental Radioactivity, 84, 333–342.

[294] Rosamilia S., Gaudino S., Sansone U., Belli M., Jeran Z., Ruisi S., Zucconi L. (2004) Uranium isotopes, metals and other elements in lichens and tree barks collected in Bosnia-Herzegovina. Journal of Atmospheric Chemistry, 49(1–3), 447–460.

[295] Daillant O., Boilley D., Gerzabek M., Porstendöerfer J., Tesch R. (2004) Metabolised tritium and radiocarbon in lichens and their use as biomonitors. Journal of Atmospheric Chemistry, 49(1–3), 329–341.

[296] Daillant O., Kirchner G., Pigree G., Porstendörfer J. (2004) Lichens as indicators of tritium and radiocarbon contamination. Science of the Total Environment, 323, 253–262.

[297] Benedini M., Cicioni G. (1992) I modelli matematici e le loro potenzialità. IRSA (Istituto di Ricerca sulle Acque), Atti della giornata di studio: Modelli

matematici per il Bacino del fiume Po, Parma, 3 June, 1992. Quad. IRSA, 95, pp. 1.1–1.30.

[298] Conti M.E. (1996) The pollution of the Adriatic Sea: scientific knowledge and policy actions. International Journal of Environment and Pollution, 6(2–3), 113–130.

[299] Janssen C.R., Heijerick D.G., De Schamphelaere K.A.C., Allen H.E. (2003) Environmental risk assessment of metals: tools for incorporating bioavailability. Environment International, 28(8), 793–800.

[300] Seaward M.R.D., Letrouit-Galinou M.A. (1991) Lichen recolonization of trees in the Jardin du Luxembourg, Paris (France). Lichenologist, 23(2), 181.F–186.F.

[301] Letrouit-Galinou M.A., Seaward M.R.D., Deruelle S. (1992) On the return of epiphytic lichens to the Jardin du Luxembourg. Bulletin – Societe Botanique de France, Lettres Botaniques, 139(2), 115–126.

[302] Bargagli R., Gasparo D., Lazzarin A., Lazzarin G., Olivieri S., Tretiach M. (1991) Lichens as indicators and monitors of atmospheric pollutants in NE Italy, preliminary data on the integrated testing system. Botanica Chronica, 10, 977–982.

[303] Garty J. (1988) Comparisons between the metal content of a transplanted lichen before and after the start-up of a coal-fired power station in Israel. Canadian Journal of Botany, 66, 668–671.

[304] Garty J., Amman K. (1987) The amounts of Ni, Cr, Zn, Pb, Cu, Fe and Mn in some lichens growing in Switzerland. Environmental and Experimental Botany, 27, 127–138.

[305] Folkeson L. (1979) Interspecies calibration of heavy metal concentrations in nine mosses and lichens: applicability to deposition measurements. Water Air and Soil Pollution, 11, 253–260.

[306] Tomassini F.D., Puckett K.J., Nieboer E., Richardson D.H.S., Grace B. (1976) Determination of copper, iron, nickel, and sulphur by X-ray fluorescence in lichens from the Mackenzie Valley, Northwest Territories, and the Sudbury district, Ontario. Canadian Journal of Botany, 54, 1591–1603.

[307] Garty J. (1985) The amounts of heavy metals in some lichens of the Negev Desert. Environmental Pollution (series B), 10, 287–300.

[308] Bargagli R., Monaci F., Borghini F., Bravi F., Agnorelli C. (2002) Mosses and lichens as biomonitors of trace metals. A comparison study on *Hypnum cupressiforme* and *Parmelia caperata* in a former mining district in Italy. Environmental Pollution (Barking, Essex: 1987), 116(2), 279–287.

[309] Puckett K.J., Finegan E.J. (1980) An analysis of the element content of lichens from the northwest territories, Canada. Canadian Journal of Botany, 58, 2073–2089.

[310] Allen-Gil S.M., Ford J., Lasorsa B.K., Monetti M., Vlasova T., Landers D.H. (2003) Heavy metal contamination in the Taimyr Peninsula, Siberian Arctic. Science of the Total Environment, 301(1–3), 119–138.

6 Biomarkers for human biomonitoring

A Alimonti & D Mattei

1 Introduction

People are continuously exposed to thousands of natural and man-made chemicals through the environment, food habits and lifestyles. Using modern analytical technology it is now possible to measure a large number of chemicals and their metabolites present in the human organism (in blood, tissues, urine, hair, etc.). The biomonitoring is a procedure well known since 1927 when the first paper on the use of the analysis of lead in urine in exposed workers was published. Today biomonitoring is largely used to control the health risk of people occupationally and non-occupationally exposed. Programmes on the biomonitoring are currently in progress in the USA [1] and in Europe [2]. The biomonitoring evaluates the human exposure by comparison with appropriate reference values and goes by the knowledge of the relationship between environmental exposure and deriving degree of adverse health effects. When a health risk is revealed, legislators may decide to ban a product or restrict its usage to applications with lower risks for human health. Biomonitoring techniques are becoming, in fact, common tools for decision-makers in the health and environmental field. Consequently, biomarkers can be considered suitable tools to measure the impact of a contaminant on an organism, characterized by its interaction with the endogenous molecules. Biomarkers are able to point out biochemical, genetic, morphological or physiological changes in the organism who suffers from a particular stress situation – due to occurrence of heavy metals, pesticides, etc. Biomarkers suggest the occurrence of toxicological events much earlier than the emergence of those effects that can be evaluated. A biomarker definition is: "… a change, produced by a contaminant, at biochemical or cellular level of a process, a structure or a function that can be measured in a biological system" [3]. This change provides information (qualitative, semi-quantitative or quantitative) about the chemical source, and on the correlation between the biological effects and the environmental contamination levels. A contaminant can cause primary toxicity at biochemical and molecular levels (alterations in enzymatic activity, DNA level, etc.), and, secondly, through cascade events can cause toxicity at cellular, tissue or organism levels. A few homoeostatic responses to a chemical damage represent some possible biomarkers that can be applied for toxicological investigations [4, 5]. Selection of appropriate biomarkers is a critical point, which depends upon the state of scientific knowledge and can be influenced by social, ethical and economic factors. Also the identification of "practicable and realistic biomarkers" associated with different toxic end-points or outcomes is a key step. It is well known that lifestyle is implicated in determining the risks for development of cancers, circulatory diseases, neurodegenerations and other chronic diseases. Similarly, life events and periods, such as childhood, reproduction and senescence, may affect the distribution of chemicals within the body. During pregnancy, as an example, many chemicals may pass the placental barrier causing exposure of the foetus. Lactation may result in excretion of lipid-soluble chemicals, thus leading to a decreased retention in the mother along with an increased uptake by the infant. During weight loss or development of osteoporosis, stored chemicals may be released, resulting in a renewed and protracted "endogenous" exposure of target organs. Other factors may affect individual absorption, metabolism, retention and distribution of chemical compounds and they have to be considered when a biomarker has to be measured.

The most important features of a biomarker are: (i) stability (to allow the biological sample preservation); (ii) sensitivity (i.e. low probability of false negative); and (iii) specificity (i.e. low probability of false positive). Moreover, a biomarker must reflect the interaction (qualitative or quantitative) of the host biological system with the compound of interest and it has to be reproducible qualitatively and quantitatively with respect to time (short- and long-term). Moreover, the biomarker use must be generally regarded as ethically acceptable. Biomarkers can be classified in: (i) biomarkers of exposure; (ii) biomarkers of effect; and (iii) biomarkers of susceptibility.

1.1   Biomarkers of exposure

Biomarker of exposure, the first kind of biomarkers used in human biomonitoring studies, may be an exogenous compound or its metabolite (i.e. a metal or a metal compound) inside the body, an interactive product between the compound (or metabolite) and an endogenous component, or another event related to the exposure. Often, there is not a clear distinction between exposure and effect biomarkers. For example, adducts formation could reflect an effect rather than the exposure. However, biomarkers of exposure usually indicate changes in the functions of the cell, tissue or total body. Usually, they comprise measurements of the compounds in appropriate samples, such as blood, serum or urine [6–15]. Volatile chemicals concentration may be assessed in exhaled breath, after inhalation of contamination-free air. Biomarkers of exposure may be used to identify exposed individuals or groups, quantify their exposure, assess their health risks, or to assist in diagnosis of diseases with environmental or occupational aetiology. The biomarkers of exposure must be evaluated also with regard to temporal variation. The degree of concentration in the blood reflects levels in different organs and varies widely between different chemicals, and usually also depends on the length of the exposure. Sometimes this type of evidence is used to classify a biomarker as a marker of (total) absorbed dose or a marker of effective dose (i.e. the amount that has reached the target tissue). For example, exposure to a particular solvent may be evaluated from data on the actual concentration of the solvent in the blood at a particular time following the exposure. This measurement will reflect the amount of the solvent that has been absorbed into the body. Some of the absorbed amount will be exhaled due to the vapour pressure of the solvent. While circulating in the blood, the solvent will interact with various components of the body, and it will eventually become subject to breakdown by enzymes. The outcome of the metabolic processes can be assessed by determining specific mercapturic acids produced by conjugation with glutathione. The cumulative excretion of mercapturic acids may better reflect the effective dose than will the blood concentration. However, biomarkers of exposure alone do not give information on the sources or levels of exposure; when, where, how or how many times the exposure occurred; or any relationships between exposure and health effects. Recent technological advances in genomics, proteomics and metabolomics are providing new tools for investigating endogenous chemicals that can be used to characterize individual's exposure to a single chemical or a mixture of chemicals.

1.2   Biomarkers of effect

Biomarkers of effect (BEF) are referred to reversible biochemical and functional alterations than can be measured in a target tissue of the organism. A BEF may be an endogenous component, or a measure of the functional capacity, or some other marker of the state or balance of the body or organ system, as affected by the exposure. Usually it is a pre-clinical marker of pathology and can be specific or non-specific [16]. Specific biomarkers are useful because they indicate a biological effect of a particular exposure, thus providing evidence that can potentially be used for preventive purposes. The non-specific biomarkers do not point to an individual cause of the effect, but they may reflect the total, integrated effect due to a mixed exposure. Both types of biomarkers of effect are useful biomarkers of early (critical) effects. For example, the detection of early damage to the kidney tubules caused by exposure to Cd using urinary levels of low molecular weight proteins such as β_2-microglobulin, protein HC and the enzyme N-acetylglucosaminidase. Technical developments have been occurred with biomarkers of effect to mutagenic chemicals. These compounds are reactive and may form adducts with macromolecules, such as proteins or DNA. DNA adducts may be detected in white blood cells or tissue biopsies, and specific DNA fragments may be excreted in the urine. Other macromolecules may also be changed by adduct formation or oxidation. In particular, such reactive compounds may generate haemoglobin adducts that can be determined as biomarkers of effect to these compounds. For the purpose of occupational health, these biomarkers should be restricted to those that indicate sub-clinical or reversible biochemical changes, such as inhibition of enzymes. The most frequently used biomarker of effect is probably the inhibition of cholinesterase caused by certain insecticides, namely, organophosphates and carbamates. In most cases, this effect is entirely reversible, and the enzyme inhibition reflects the total exposure to this particular group of insecticides. Some exposures do not result in enzyme inhibition but in increased activity of an enzyme. This is the case of several enzymes belonging to the P450 family. They may be induced by exposures to certain solvents and polyaromatic hydrocarbons. Generally, the enzyme activity is determined indirectly *in vivo* by managing a compound that is metabolized by that particular enzyme, and then the breakdown product is measured in urine or plasma. Other exposures may induce the synthesis of a protective protein in the body. The best example is probably metallothionein, which binds cadmium (Cd) and promotes its excretion; Cd exposure is one of the factors resulting in increased expression of the metallothionein gene. Similar protective proteins may exist but have not yet been explored sufficiently to become accepted as biomarkers. Among the candidates for possible use as biomarkers are the so-called stress proteins, previously known as heat shock proteins. These proteins are generated by a range of different organisms in response to a variety of adverse exposures. The urinary excretion of proteins with a small molecular weight, such as albumin, may be used as a biomarker of early kidney damage. Relating to genotoxic effects, chromosomal aberrations or formation of *micronuclei* can be detected by microscope observation. Damage may also be revealed by adding a dye to the cells during cell division. Exposure to a genotoxic agent can then be visualized as an increased exchange of the dye between the two chromatids of each chromosome (sister chromatid exchange, SCE). Chromosomal aberrations are related to an increased risk of developing cancer. More sophisticated assessment of genotoxicity is based on particular point mutations in somatic cells, that is, white blood cells or epithelial cells obtained from the oral mucosa.

1.3 Biomarkers of susceptibility

Biomarkers of susceptibility are indices of the individual predisposition (hereditary or acquired) to suffer xenobiotic effects, that is, to be particularly sensitive to the effects of a single compound or of a group of such chemicals. On the contrary of the genes associated with hereditary pathologies, the genes of susceptibility are able to alter the risk for an adverse effect in case of exposure to dangerous chemical compounds [17]. Although other factors may be important, the ability to metabolize certain chemicals is considerably variable and is genetically determined; therefore, most attention has been recently focused on genetic susceptibility. Several relevant enzymes appear to be controlled by a single gene. For example, oxidation of foreign chemicals is mainly carried out by a family of enzymes belonging to the P450 family. Other enzymes make the metabolites more water soluble by conjugation (e.g. *N*-acetyltransferase and *m*-glutathion-*S*-transferase). Important studies suggest that a risk of developing certain cancer forms is related to the capability of metabolizing foreign compounds. Many questions still remain unanswered, limiting the use of these potential biomarkers of susceptibility, both in occupational medicine and in public health. Individuals with a chronic disease may be more sensitive to an occupational exposure. Also, if a disease process or previous exposure to toxic chemicals has caused some sub-clinical organ damage, then the capacity to withstand a new toxic exposure is likely to be less. Biochemical markers of organ function may in this case be used as biomarker of susceptibility. Perhaps the best example regarding hypersusceptibility relates to allergic responses. If an individual has become sensitized to a particular exposure, then specific antibodies can be detected in serum. Even if the individual has not become sensitized, other current or past exposures may add to the risk of developing an adverse effect related to an occupational exposure. A major problem is to determine the joint effect of mixed exposures at work. In addition, personal habits and drug use may result in an increased susceptibility. For example, tobacco smoke usually contains a considerable amount of Cd. Thus, with occupational exposure to this element, a heavy smoker who has accumulated substantial amounts of Cd in the body will be at increased risk of developing Cd-related kidney disease.

2   Chemical elements

2.1   Aluminium

Aluminium (Al) is not widely accumulated in plants or animals, with some exceptions, such as tea plant, some mosses, ferns and subtropical evergreen trees. Aluminium does not appear to accumulate to significant level in cow's milk or beef tissue and is not expected to biomagnificate. The daily ingestion of Al by humans is estimated to be 30 to 50 mg [18]. Small amounts of ingested Al reach the bloodstream and are then excreted by urine; while the largest quantity is quickly excreted by faeces. Moreover, the Al bioavailability is strongly influenced by its compound and the presence of dietary constituents which can complex the metal and thereby enhance or inhibit its absorption, i.e. diets with low levels of Cu daily intake result in increased deposition of Al in body tissues [19]. The main mechanism of absorption is probably passive diffusion through paracellular pathways, while the dermal absorption is not the main way of exposure. In foundry workers [20], welders [21] and workers exposed to fine Al dust [22] wheezing, dyspnoea, and impaired lung function have been observed. Pulmonary fibrosis is the most commonly reported respiratory effect observed in workers exposed to fine Al dust, alumina or bauxite. However, conflicting reports are available on the fibrogenic potential of Al. Al toxicity in general population is known to occur in at least two specific situations: neurological diseases (including amyotrophic lateral sclerosis, dementia associated with Parkinson's disease and Alzheimer's disease) [23] and osteomalacia or metabolic bone disease (in dialysis patients) [24, 25]. Cerebral disfunction was reported in people exposed to drinking water that had been contaminated with Al sulphate [26].

2.1.1 Biomarkers of exposure

Aluminium can be measured in the blood, bone, urine and faeces. Unfortunately, exposure levels cannot be related to serum or urine levels very accurately, primarily because Al is very poorly absorbed by any route and its oral absorption in particular can be quite affected by other concurrent intakes. There is an indication that high exposure levels are reflected in urine levels, but this cannot be well quantified as much of the Al may be rapidly excreted. Aluminium can also be measured in the faeces, but this cannot be used to estimate absorption. Despite the preceding observations, urinary levels for Al (U-Al) has been set by the German Research Foundation (Deutsche Forschungsgemeinschaft, known as DFG). The DFG established a Biological Tolerance Values (Biologische Arbeitsstofftoleranzwerte, known as BAT) for U-Al of 200 µg/L, while the reference range for the general population is 1–15 µg/L.

2.1.2 Biomarkers of effect

There are no known simple, non-invasive tests which can be used as biomarkers of effect caused by Al. D'Haese et al. [27] proposed the use of the desferrioxamine test to identify individuals with Al-related bone disease/Al overload. This test involves administering a challenge dose of the chelator desferrioxamine to individuals with suspected Al-induced bone disease. However, iron supplementation may interfere with the test results [28].

2.1.3 Biomarkers of susceptibility

No biomarkers of susceptibility are available at the moment.

2.2 Arsenic

Cigarette smoke is an important source for inhaled arsenic (As), but food is the largest source of the element for general population. Seafood contains great amounts of As, tough fish and shellfish contain arsenobetaine, an As organic form less toxic than the other As compounds. Daily dietary intake to arsenic is highly variable, with a mean of 50.6 µg/day for females and 58.5 µg/day for males. Occupational exposure to As includes workers involved in pesticides and wood treating industries. Most cases of As-induced toxicity in humans are due to exposure to inorganic As, but some recent animal studies suggest not to discharge also the possibility of health effects from the organic As forms [29]. The International Agency for Research on Cancer (IARC) classified inorganic As compounds as carcinogens to human (Group 1), in fact As has been recognized to increase the risk for liver, bladder, kidneys, prostate and lungs cancer. Also the US Environmental Protection Agency (EPA) determined that inorganic As is a human carcinogen by the inhalation and oral route (Group A). Under discussion are the effects caused by As to foetus [30]. Arsenic compounds are generally metabolized in the liver where they are converted in less harmful organic compounds. For example, As^{3+} can be oxidized to As^{5+} and metabolized in mono- and di-methyl arsenic acid. It is estimated that more than 75% of the absorbed As dose is excreted in urine [31], although this may vary with the dose and exposure duration. Smaller amounts are excreted in faeces. Reduced inorganic As may react with sulphydryl groups in proteins and inactivate target enzymes. A particular target in the cell is the mitochondrion, which accumulates the element [32]. Arsenic inhibits succinic dehydrogenase activity and can uncouple oxidative phosphorylation; the resulting fall in ATP levels affects all cellular functions (Na^+/K^+ balance, protein synthesis, etc.). Mechanistic studies of As toxicity seem to suggest its role in the reactive oxygen species generation [33]. Absorption across the lungs involves the deposition of the particles onto the lung surface and the absorption of As from the deposited material. Workers exposed to As trioxide dusts in smelters excreted in the urine about 40–60% of the estimated inhaled dose [34]. Furthermore, several studies in humans indicate that arsenates and arsenites are well absorbed across the gastrointestinal tract. No quantitative studies were located on absorption of inorganic arsenicals in humans after dermal exposure. On the other hand, the organic As detoxification occurs mainly through methylation reactions.

2.2.1 Biomarkers of exposure

Normal levels of As in hair and nails are reported as 1 µg/L [35]. This value may increase up to 100-fold after an arsenic exposure [36] and remain elevated for 6–12 months. But the minimum exposure level that produce measurable increases in As levels in hair and nails have not been precisely defined; for this reason hair and nails are normally not used as biomarkers for occupational exposure. Non-exposed individuals show blood As levels less than 1 µg/L [37]. Due to the quick clearance of As from blood within a few hours [34], measurements of blood As reflect only recent exposures. The blood levels, thus, do not appear to be reliable markers of chronic exposure, mainly to low levels of As. On the other hand As, absorbed from the lungs or the gastrointestinal tract, is excreted in urine within 1–2 days. For this reason, measurement of urinary As (U-As) levels is generally accepted as the most reliable marker of recent arsenic exposure. In non-exposed individuals a reference range for U-As is 2–25 µg/L. The occupational limit set by the American Conference of Governmental Industrial Hygienists (ACGIH) (2002) for U-As is 35 µg/L (Biological Exposure Indices, BEI).

2.2.2 Biomarkers of effect

The characteristic skin changes, as hyperpigmentation and hyperkeratinization, caused by As are the most sensitive and diagnostic clinical markers of chronic exposure to As. Another effect of As exposure is the peripheral neuropathy, for that reason researchers investigated the decreasing in nerve conduction velocity as a biomarker for peripheral neuropathy [38]. But the above-mentioned biomarkers cannot be used in occupational medicine because they are not specific for As exposures. Arsenic is known to influence the activity of some enzymes that can be used as biomarkers of effect: for example enzymes responsible for heme synthesis and degradation – including inhibition of coproporphyrinogen oxidase and heme synthetase [39], and activation of heme oxygenase [40]. Furthermore, Menzel *et al.* [41] reported the induction by As of human lymphocyte heme oxygenase *in vitro* and suggested this enzyme as possible biomarker of arsenite exposure.

2.2.3 Biomarkers of susceptibility

Since the degree of arsenic toxicity may be influenced by the rate and extent of its methylation in the liver it seems likely that some members of the population might be especially susceptible because of lower than normal methylating capacity. Moreover, several studies were carried out for fixing biomarkers of susceptibility, included genetic polymorphisms of enzymes involved in xenobiotic metabolism, DNA repair and oxidative stress, as well as serum level of carotenoids, but it was not still identified real human biomarkers of susceptibility for arsenic.

2.3 Beryllium

Beryllium (Be) as a chemical component occurs naturally in soil; however, industrial wastes may increase the amount of Be in soil. Furthermore, Be occurs in tobacco and may be inhaled from cigarette smoke. General population can be exposed to normal levels of Be by breathing air, eating food or drinking water that contains Be. Beryllium, if greater than 1 mg/m^3 in air, can damage lungs by inhalation similarly to pneumonia with reddening and expanding of the lungs – condition known as acute Be disease. A lot of people can develop hypersensitivity or allergy to this metal giving rise to a chronic inflammatory reaction called granulomas, characterized by breathing difficulty. Long periods of exposure to Be have been reported to initiate cancer in laboratory animals and occupational studies reported an increased risk of lung cancer. In fact, the IARC, classified Be and its compounds as human carcinogens. Based on serum levels of Be in workers accidentally exposed, the biological half-life was estimated to be 2–8 weeks [42]. Beryllium and its compounds are mainly absorbed from lungs and poorly from the gastrointestinal tract; its absorption through intact skin is questionable. Skin ulceration in workers exposed to Be occurred only after the skin was accidentally cut or abraded [43]. In lungs, Be, acting as a hapten, interacts with alveolar macrophages resulting in a Be–peptide associated with a histocompatibility class II Be–peptide complex [44], which is recognized by the T cell receptor, with the help of CD4+ molecules. This interaction activates CD4+ and T lymphocytes. The antigen-specific inflammatory response to Be is a cell-mediated process arranged by cytokines. In fact, results of several studies support the role of cytokines in chronic Be disease [45]. Absorbed Be (especially by lungs) is distributed throughout the body via blood, with the highest concentrations in the liver and skeleton [45–49]. Beryllium has been also detected in breast milk [50], although the relationship between Be exposure and breast milk levels has not been established. The primary routes of elimination of absorbed Be are urine and faeces.

2.3.1   Biomarkers of exposure

Among the markers of the body burden of Be are included its levels in the urine, blood, skin or lungs. These levels are rarely measured because they are not reliable measures of the exposure over time. Normal levels of 1 ppb for blood [42], 0.28–1 μg/L for urine [51] and of 0.02 ppm for lung tissue [52] have been reported. A sampling of 500 urine samples collected during the National Health and Nutrition Examination Survey (NHANES) III study, suggested that background levels of Be in the urine may be lower. Nonetheless, urinary excretion of Be is irregular and not useful for diagnostic purposes [53]. The level of beryllium in blood, serum or plasma is predictive of the intensity of current exposure from some Be compounds [54]. Biopsies have been analysed to determine beryllium concentrations in the body. Lung tissue of two employees of a Be extraction and processing plant, where beryllium concentrations exceeded the recommended standards of 2 μg/m^3 for an 8 h day and 25 μg/m^3 for a 30 min maximum level, contained 0.18 and 0.65 μg/g dry weight compared to the normal level of 0.02 μg/g [52]. Beryllium levels in lung biopsies indicate exposure to Be but may not confirm the presence of chronic Be disease(CBeD); neither indicates how recently the exposure occurred because the clearance of the metal from the lungs depends upon the solubility of its compound. Biomarkers of oral or dermal exposure to Be were not located, probably because very little amount of the metal is absorbed after exposure by these routes.

2.3.2   Biomarkers of effect

Because CBeD is caused by an immune reaction to the metal, several tests have been developed to assess Be hypersensitivity. The most commonly used test is the Be lymphocyte proliferation test; both peripheral blood cells and bronchioalveolar lavage cells can be used in the test, but the blood was not widely used because its difficulty to be performed and its scarce reproducibility.

2.3.3   Biomarkers of susceptibility

Experimental data suggest the genetic susceptibility to develop CBeD. A human leukocyte antigen (HLA) class II marker has been strongly associated with CBeD. Studies conducted by Fontenot et al. [55] and Lombardi et al. [56] suggest that the HLA-DP alleles, the human leukocyte antigen system (HLA) class II (DP), in particular those with HLA-DP containing Glutamate69 (Glu69), are involved in the presentation of Be to CD4+ T cells, which are implicated in the pathogenesis of chronic Be disease. Richeldi et al. [57, 58] found a higher frequency of allelic variants of the HLA-DP gene coding for a glutamate in position 69 of HLA-DPB1 chain (HLA-DPB1 Glu69) among individuals with chronic Be disease than in Be-exposed individuals without the disease. The HLA-DPB1 Glu69 DNA marker was found in 5 of 6 workers with chronic disease, 0 of 2 Be-sensitized individuals without disease and 36 of 119 (30%) unsensitized Be-exposed individuals [57, 58].

2.4 Cadmium

About 25,000 to 30,000 tons of Cd are released each year to the environment and, then, to the oceans; half deriving from the weathering of rocks into rivers' water. Release of Cd from human activities is estimated at from 4,000 to 13,000 tons per year. Grain and leafy vegetables, which readily absorb Cd from the soil and fish, that accumulated it in their edible parts, are other sources of Cd. The metal is also found in cigarette fumes (0.007–0.35 µg per cigarette) and exhaust from vehicles [59]. After absorption, Cd is transported in the blood bound to albumin with a half-life up to 3 months. It is taken up by the liver, and, due to its similarity to zinc, causes this organ to induce the synthesis of the protein metallothionein. The Cd–metallothionein (Cd–MT) complex then becomes transported to the kidneys, and it is filtered at the glomerulus, but is reabsorbed at the proximal tubule. Within the renal tubular cells, the Cd-MT complex becomes degraded by digestive enzymes, which releases the Cd. Renal tubular cells deal with the release of this toxic substance by synthesizing MT to neutralize it, but eventually the kidneys loose their capacity for producing MT. At this point, the Cd has accumulated to a high level in the renal tubular cells, and an irreversible cell damage occurs. But the renal cells do not have an effective elimination pathway for the Cd complex, which implicates that the half-life in the kidney is 15–30 years. Cadmium, like similar heavy metals, can inactivate enzymes containing sulphydryl groups and it can also compete with other metals (i.e. Zn and Se) for inclusion into metallo-enzymes. Cadmium can also compete with Ca for binding sites on regulatory proteins such as calmodulin [59]. An acute intake of Cd causes testicular damage with degeneration of the testes and decreasing in spermatozoa. Furthermore, if Cd is inhaled, lung irritation can occur. Long-term ingestion causes kidney damage (i.e. proteinuria and β_2-microglobulinuria) and disorders of Ca metabolism occur, causing osteomalacia. The metal is also known to be carcinogenic, Group 1 by IARC [60], and in studies has been linked with cancers in the lungs and prostate. The U.S. Occupational Safety and Health Administration (OSHA) sets these permissible exposure limits (PELs) of 0.1 mg/m^3 as an 8 h time-weighted average (TWA) for fumes and 0.2 mg/m^3 as TWA for dust.

2.4.1 Biomarkers of exposure

Cadmium levels in blood, urine, faeces, liver, kidney, hair and other tissues have been used as biological markers of exposure. Blood Cd (B-Cd) levels are principally indicative of recent exposure to the metal [61]. Concentrations of B-Cd in normal populations range from about 0.4 to 1.0 µg/L for non-smokers and 1.4–4.0 µg/L for smokers [62]. Workers occupationally exposed to Cd by inhalation may have B-Cd levels up to 50 µg/L, but B-Cd contents up to 10 µg/L are considered acceptable in occupational exposures [63]. Urinary Cd (U-Cd) reflects the total body burden but responds only to recent exposure [64]. In non-smokers children, the U-Cd reference value is about 0.31–0.45 µg/L [65]. In populations with substantial environmental or occupational exposure, values can range up to 50 µg/g creatinine [66], with interval concentrations from 0.1 to 4.0 µg/L.

Liver and kidney tissues preferentially accumulate Cd with age and degree of exposure. Hair levels of Cd have been used as a measure of Cd exposure, although the possibility of exogenous contamination has led to substantial controversy concerning the reliability of hair levels as a measure of absorbed dose [64]. In conclusions measurement of Cd levels in various biological materials can provide an indication of recent or total Cd exposure, but the probability of adverse effects cannot be reliably predicted except at high exposure levels.

2.4.2   Biomarkers of effect

Acute inhalation exposure to high levels of Cd causes respiratory damage and may lead to death. No information was located on biomarkers of respiratory effects in humans, but based on animal experiments, activity of alkaline phosphatase in the surfactant fraction of bronchial-alveolar fluid suggested as a marker of pulmonary damage after an acute Cd inhalation [67]. Renal dysfunction is considered the main toxic effect of chronic Cd exposure. Impaired kidney function has been measured by increased levels of solutes (proteins, amino acids, uric acid, calcium, copper, phosphorous, etc.) in urine and/or serum. For example, urinary β_2-microglobulin is used as a marker of tubular renal dysfunction [68]. Furthermore, urinary N-acetyl-β-D-glucosaminidase (NAG), a lysosomal enzyme present in high concentrations in the proximal tubule, has a better correlation with urinary Cd levels than does β_2-microglobulin at low Cd exposure levels [69–70]. At present, not enough information is available to determine which, if any, of these parameters provide sensitive and specific markers of Cd-induced renal damage.

2.4.3   Biomarkers of susceptibility

No reliable biomarkers of susceptibility are available at the moment, although Lu *et al.* [71] reported that MT gene expression in peripheral blood lymphocytes can be used as a biomarker of Cd exposure and susceptibility to renal dysfunction.

2.5   Chromium

The primary route of chromium (Cr) absorption is inhalation of Cr dust. Cr (VI) compounds are more readily absorbed from the lungs than Cr (III) compounds, due in part to differences in the capacity to penetrate biological membranes. Studies have shown that women have lower Cr deposits in various tissues (especially in lungs) than those found in men; this can reflect the greater potential for occupational exposure in men. In the stomach, Cr (VI) compounds are reduced to Cr (III) compounds, explaining the relatively poor gastrointestinal absorption of orally administered Cr (VI) compounds. Systemic toxicity has been observed in humans following dermal exposure to Cr compounds. In these cases the major manifestations are ulcers (chrome holes). In general, toxicity of Cr is mainly caused by hexavalent compounds. This is due to the chromate anion $(CrO_4)^{2-}$ that can enter the cells through non-specific anion channels, while absorption of Cr (III) compounds is via passive diffusion and phagocytosis. Chromium (III) is bound to plasma proteins such as transferrin, while Cr (VI) enters in the blood stream, then is taken up and reduced by erythrocytes (by many substances including ascorbate and glutathione) and at least it is found bound to haemoglobin. Chromium (VI) is a strong clastogen and produces chromosome aberrations and SCEs as well as DNA strand breaks, oxidized base damage, DNA–DNA and DNA–protein crosslinks. The IARC [72] classified Cr (VI) in Group 1 – carcinogenic to humans – based on evidence in humans for the carcinogenicity of Cr (VI) compounds (increased risk of respiratory cancers, primarily lung and naso-pharyngeal). Chromium is normally excreted through the kidneys in urine, with some excretion through the bile and faeces; minor routes of excretion include breast milk, sweat, hair and nails. Workers exposed mainly to Cr (VI) compounds had higher urinary Cr concentrations than workers exposed primarily to Cr (III) compounds. However, Cr (VI) is not detectable in urine, because it is rapidly reduced before excretion.

2.5.1 Biomarkers of exposure
Urinary Cr is a well-known biomarker of exposure. The levels generally found in people not occupationally exposed are ≤ 2.0 µg/L, while limit levels for occupational exposure are (as BEI) 10 µg/g creatinine during the 8 h workshift as total Cr (30 µg/g creatinine at the end of the week work) [73]. Blood Cr is another reliable biomarker with reference value in general population <0.75 µg/L. Occupational limits in blood set by DFG are: 9 µg/L corresponding to 0.3 mg/m^3 (as environmental Cr (VI)); 17 µg/L corresponding to 0.05 mg/m^3; 25 µg/L corresponding to 0.08 mg/m^3; and 35 µg/L corresponding to 0.10 mg/m^3 [74].

2.5.2 Biomarkers of effect
There are no biomarkers of effect set at the moment, but it has been observed that the presence of low molecular weight proteins, such as retinol binding protein, antigens or β_2-microglobulin in the urine, could be an early indication of kidney dysfunction. Furthermore some cell culture studies demonstrated that Cr forms protein–DNA crosslinks and adducts with DNA – indicating potential genotoxic effects or cancer in humans and that these end points may be potentially useful biological markers [75].

2.5.3 Biomarkers of susceptibility
No biomarkers of susceptibility are at the moment suggested.

2.6 Cobalt

Cobalt (Co) may be released into the atmosphere from coal-fired power plants and incinerators, vehicular exhaust, industrial activities relating to the mining and processing of Co-containing ores, and the production and use of Co alloys and chemicals. ^{58}Co and ^{60}Co isotopes may be released to the environment as a result of nuclear accidents and radioactive waste dumping. Observations on workers have shown that Co is absorbed by inhalation and ingestion. Inhaled Co particles, depending on the particle size, can deposit in the upper and lower respiratory tract and Co is subsequently absorbed. Gastrointestinal absorption of Co is influenced by the type and dose of Co compound given and by nutritional factors. It is reported to vary from 5–45% in humans. Dermal exposure can be considered negligible in comparison with other kind of exposures, but allergic dermatitis, photosensitization (e.g. in cement) and cross-sensitization with nickel (Ni) have been observed. Cobalt has an erythropoietic effect and has been used for the treatment of anaemia. No data on genotoxicity were available by IARC but it has determined that Co is possibly carcinogenic to humans. Cobalt is also essential as a component of vitamin B12; therefore, it is physiologically found in most tissues, in particular liver and the kidney. The total amount of Co stored in the body is around 1.5 mg and it does not seem to be age-dependent. The half-life of Co in the whole body ranges from approximately 5 days to 4 years, depending on the compound and the route of exposure. It is mainly excreted in urine and moderately *via* faeces. The urinary elimination is characterized by a rapid phase of a few days duration followed by a second phase which may last a couple of years. In non-occupationally exposed subjects, the concentrations of urinary and serum Co are usually <2.0 µg/g creatinine and 0.4 µg/L, respectively [76]. It has been reported that the measurement of urine or blood Co at the end of the workweek could be recommended for the assessment of recent exposure in the case of exposure to soluble Co compounds. An 8 h exposure to 20 or 50 µg/m³ of a soluble form of Co would lead to an average concentration in urine sample collected at the end of the workweek of 18.2 or 32.4 µg/g creatinine, respectively [76].

2.6.1   Biomarkers of exposure
Positive correlation between occupational exposure to Co and its levels in both urine and blood is assessed [77, 78]. The highest excretion rate of Co in urine occurs during the first 24 h after short-term exposure. Occupational exposure to 0.1 mg/m³ Co resulted in urinary levels from 59 to 78 µg/L [76]. Some Italian reference values for serum Co is 0.19 ± 0.11 µg/L and for blood Co is 0.12 ± 0.08 µg/L [79].

2.6.2   Biomarkers of effect
There are no established biomarkers of effect till now for Co, but laboratory studies reported that sensitive serum protein responses were found in animals exposed to Co at levels below those necessary to produce haematopoietic effects [80]. These serum protein responses included an increase in α-globulin fractions and associated serum neuraminic acid. Sensitization to Co results in Co-specific changes in serum antibodies (IgE and IgA) that may be monitored to determine if sensitization or additional exposure to Co has occurred [81].

2.6.3   Biomarkers of susceptibility
Potolicchio *et al.* [82] have suggested that individuals with a polymorphism in the HLA-DP gene (presence of Glutamate 69 in the *β* chain) may be more susceptible to hard metal lung disease, but there are no specific biomarkers of susceptibility for Co at the moment.

2.7 Lead

Toxic effects of lead (Pb) on humans are well documented. Lead poisoning has been reported worldwide, although in many countries Pb exposure is decreased since 1980s especially for the increasing use of unleaded gasoline. Mortality is rare today. However, death during the 1960s from Pb encephalopathy was not rare in urban centres. Children are at the greatest risk for Pb poisoning, because they are most likely to put things containing Pb into their mouths, because their brains are rapidly developing and are most vulnerable to any disorganizing influence, and because they absorb up to 50% of the amount of Pb ingested. Perhaps the best-known effects are the interferences with the critical phases of the dehydration of aminolevulinic acid and the incorporation of Fe into the protoporphyrin molecule; the result is a decreased heme production. Because heme is essential for cellular oxidation, deficiencies have extensive effects [83]. Due to the fact that Pb is an enzymatic poison, it perturbs multiple essential bodily functions, producing a wide array of symptoms and signs (on brain, kidney, bone, etc.) [84]. Data of epidemiological studies in workers reported excess of deaths from bladder, stomach and lung cancer [85]. The carcinogenicity of lead was recently evaluated by IARC [86] and, although the absence of an adequate dose–response pattern in some epidemiological studies was recognized, the association between lung, stomach cancer and lead and weak association between kidney, brain cancer and lead was confirmed (group 2B, classified as possible carcinogen for humans). Elemental and inorganic Pb compounds are absorbed through ingestion or inhalation. Pulmonary absorption is efficient, particularly if the Pb particle diameters are 1 μm generally are completely absorbed by the *alveoli*. Gastrointestinal absorption of Pb is less effective and may increase with iron, zinc and calcium deficiencies. Dermal absorption is minimal, especially for inorganic Pb compounds. Cigarettes, tobacco and alcoholic beverages are other sources of inorganic Pb. Lead is absorbed into blood plasma, crosses membranes such as the blood-brain barrier and the placenta, and accumulates in soft and hard tissues. In the blood, around 95 to 99% of Pb is sequestered in red cells, where it is bound to haemoglobin and other components. The half-life of Pb differs for each of the compartments, ranging from less than 1 h in plasma, 25–40 days in erythrocytes, 40 days in soft tissues and as many as 28 years in bone. Bone Pb accounts for more than 95% of the Pb burden in adults and 70% of the burden in children. Lead is commonly incorporated into rapidly growing bones and it competes with calcium and may exert toxic effects on skeletal growth. Lead that is not retained by the body is excreted unchanged in urine (65–75%) and in bile (25–30%). Lead also appears in hair, nails, sweat, saliva and breast milk. A biological limit of 300 μg/L has been suggested for blood Pb (B-Pb) in males and females by the Scientific Committee on Occupational Exposure Limits [87]. According to ACGIH [88] long-term exposure with <400 μg/L B-Pb should be sufficient to reduce health risks. In Germany, the Maximum Allowable Concentration (Arbeitsplatz-Konzentration known as MAK value) for Pb and its compounds is 0.1 mg/m^3 and the correlated BAT/BEI value is 400 μg/L.

2.7.1   Biomarkers of exposure

Blood-Pb (B-Pb) level is currently regarded as the most reliable index of exposure, especially at low levels [89, 90]. In general population B-Pb levels are widely distributed; for example a study conducted in Germany revealed that the geometric mean of Pb in children amounted to 31 µg/L [91]; on the other hand, these levels are higher in other countries [90]. In Italy, B-Pb as biomarker of exposure is currently used and a reference value of 60 µg/100 mL (40 µg/100 mL in fertile women) has been set by the legislative decree 25-02-2002. However, nonlinear relationships between uptake and B-Pb and between metabolic/toxic effects and B-Pb have been observed, due to the saturation of the erythrocyte lead content. A few years ago, the plasma-Pb level was proposed as alternative biomarker. Furthermore, urinary Pb is widely utilized as a biomarker of Pb exposure. The bone-Pb levels, determined by X-ray fluorescence techniques since 1970s, raises with time of employment up to 100 µm/g so that it can be useful to evaluate a long-term Pb exposure.

2.7.2   Biomarkers of effect

The decreasing in enzyme activity in the biochemical pathways in critical organs – such as bone marrow, central and peripheral nervous system, kidney and digestive system – can be used as biomarkers of effects. Inhibition of δ-aminolevulinic acid dehydratase (ALAD) and the variation in some metabolites concentration (e.g. δ-aminolevulinic acid in urine, blood or plasma, coproporphyrin in urine, zinc protoporphyrin in blood) can be also useful biomarkers of effect. This is due to the fact that most lead poisoning symptoms occur by interfering with ALAD, a zinc-binding protein implicated in the biosynthesis of heme. Also the decreasing activity of pyrimidine nucleotidase in blood and the altered nucleotide contents in blood mirror Pb exposure.

2.7.3   Biomarkers of susceptibility

In Pb exposure ALAD polymorphism – ALAD gene located in chromosome 9q34 – is related in susceptibility of Pb effect on heme metabolism. ALAD catalyses the asymmetric addition of two molecules of aminolevulinic acid to form porphobilinogen in the second step of heme synthesis. Eight ALAD variants have been described [92]. Recently, several groups have investigated on the relationships between ALAD polymorphism and susceptibility to Pb toxicity. Particular attention has been set to the G177C polymorphism, which yields two alleles, designated ALAD-1 and ALAD-2. The prevalence of the ALAD-2 allele varies from 0 to 20% depending on the population [92].

2.8   Mercury

In the environment mercury (Hg) occurs in metallic form, as inorganic Hg and organic Hg. Generally, environmental levels of Hg are quite low, for instance, between 10 and 20 ng/m^3 of Hg have been measured in urban outdoor air (i.e. hundreds of times lower than safe levels to breathe) or less than 5 ng/L in surface waters (i.e. about a thousand times lower than safe drinking water). A potential source of exposure is Hg released from dental amalgam fillings, which can contain approximately 50% of metallic Hg. Some people may be exposed to high levels of methyl-Hg if they eat often fish, shellfish or marine mammals. Hg is not essential to living cells; its absorption distribution and biotransformation are influenced significantly by its valence state. Inhalation of elementalmercury Hg° vapour is associated with an acute, corrosive bronchitis or pneumonitis. Chronic exposure to Hg vapour results in toxicity of the central nervous system including tremors, increased excitability and delirium. Elemental Hg is eventually oxidized to Hg (II) in the body by the hydrogen peroxidase-catalase pathway and is primarily excreted *via* the kidneys. However, a small portion may be exhaled. Ingestion of inorganic, oxidized Hg can result in abdominal cramping, ulceration and renal toxicity. Hg has a strong affinity for sulphur, and Hg primary mode of toxic action in living organisms is thought to be the interference of enzyme function and protein synthesis by binding to sulphydryl or thiol groups. Excretion by kidneys is the primary route of elimination of oxidized Hg, and because of its strong affinity for protein, proteinuria is a symptom associated with exposure to Hg(II). Organic Hg is highly lipophilic and exposure occurs primarily via consumption of contaminated fish. Both methyl-Hg and Hg° cross the placental (inducing teratogenic effects) and blood-brain barrier where they can be oxidized and accumulated. Methyl Hg can react directly with important receptors in the nervous system, such as the acetylcholine receptors in the peripheral nerves. Carcinogenicity and mutagenicity are not commonly associated with Hg exposure. Instead, the IARC [93] have not classified Hg as human carcinogenicity, while EPA [94] has determined that Hg chloride and methyl-Hg are possible human carcinogens.

2.8.1 Biomarkers of exposure

Blood and urine Hg (B-Hg, U-Hg) concentrations are commonly used as biomarkers of exposure. U-Hg is a biomarker used for detecting elemental and inorganic forms of Hg. The reference value in urine is less than 7 µg/L, while occupational limits are a BEI of 35 µg/g creatinine [87] and a BAT of 100 µg/L [95]. However, B-Hg levels are used to, with a reference value for the total inorganic less than 5 µg/L; the International Commission on Occupational Health (ICOH) and the International Union of Pure and Applied Chemistry (IUPAC) Commission on Toxicology determined that a mean value of 2 µg/L was the background blood level in persons who do not eat fish [96]. Occupational limits are a BEI of 15 µg/L [87] and a BAT of 25 µg/L [95]. But B-Hg levels peak quickly soon after short-term exposures, so measurements should be made soon after exposure [97]. Furthermore, there are several studies in which hair have been used as a biomarker of exposure to methyl-Hg. Once Hg is incorporated into hair, it remains unchanged [98]. A strong correlation has been found among the amount of fish swallowed, the Hg fish level and the Hg hair level. Expired air samples have been considered as possible biomarkers of exposure for Hg, but results showed that expired air can only be used soon after short-term exposure to Hg vapours.

2.8.2   Biomarkers of effect

A number of possible biomarkers of effect have been investigated, especially for neurological and renal dysfunctions. For example, the toxic effects observed at kidney level have been well correlated with blood and urine levels [99]. Biomarkers for decreased kidneys function include increase in urinary proteins and elevation of serum creatinine or β_2-microglobulin. Biomarkers for biochemical changes include eicosanoids, fibronectin, kallikrein activity and glycosaminoglycans in urine. Glomerular changes have been reported as increased high-molecular weight proteinuria [100]. Tubular changes in workers include an increasing urinary excretion of NAG, β-galactosidase and retinol binding protein [101]. But toxic kidneys parameters are not specific markers for Hg exposure and may be a consequence of other concurrent chemical exposures and they cannot be assessed as biomarker of effect till now. The neurophysiological and neuropsychological health effects of Hg have been extensively studied in occupationally exposed individuals. Neurological changes induced by Hg may look like exposure to other chemicals that can cause damage to the brain. Some studies have examined the relationship between nerve function and Hg levels in blood, urine and tissue. Tissue levels of Hg have also been found to correlate with impaired nerve function. But also in this case no biomarkers of effect are established. Potential biomarkers for the autoimmune effects of mercury have been examined and they include measurement of antiglomerular basement membrane antibodies, anti-DNA antibodies, serum IgE complexes and total IgE [99].

2.8.3   Biomarkers of susceptibility

Various factors affect the absorption, distribution, biotransformation, excretion and, consequently, toxicity of Hg. In the case of methyl-Hg, reduced glutathione and γ-glutamyl transpeptidase are involved in the excretion of methyl-Hg [102]. The glutathione S-transferase (GST) gene family is involved in the detoxification of electrophilic compounds by conjugation and a study conducted by Brambila *et al.* [103] showed that various GST genes are activated in rats exposed to Hg, indicating that individuals with specific genotypes could be better protected against the cytotoxicity of that compound.

2.9   Nickel

Food is the major source of Ni exposure, but breathing air, drinking water and smoking tobacco are too. Furthermore, jewels coated with Ni or made from Ni alloys and artificial body parts made from Ni-containing alloys are other important ways of exposure to the metal. Soluble compounds are more readily absorbed from the respiratory tract. About 20–35% of the inhaled less-soluble Ni that is retained in the lungs is absorbed into the blood. The remainder is either swallowed, expectorated or remains in the respiratory tract. Ni carbonyl is the most acutely toxic compound, causing nausea, headache, vertigo, upper airway irritation and substernal pain, followed by interstitial pneumonitis with dyspnoea and cyanosis. Dermal absorption of Ni is quantitatively minor, but it may be important in the pathogenesis of contact dermatitis caused by Ni hypersensitivity. Water soluble Ni salts are generally not carcinogenic, probably because they cannot enter cells. Water insoluble Ni compounds can produce their carcinogenic response because they are actively phagocytized into the cells. For such chemical compounds epidemiological studies have demonstrated a strong association between lung and nasal cancers and Ni exposure by inhalation in Ni refineries. The levels of Ni in biological fluids, hair and some other materials increase remarkably in persons with increased occupational or environmental exposure and decline rapidly when exposure is reduced or stopped. Seventy-five percent of plasma Ni is carried by the circulating proteins, e.g. albumin, α_2-macroglubulin and nickeloplasmin. As Ni^{2+} ions have no specific uptake mechanism, they probably enter the cells by way of Ca^{2+}/Mg^{2+} channels. Urinary excretion is the major route for the Ni elimination. Faecal excretion primarily reflects the metal that is unabsorbed from the diet and passes through the gut. All body secretions appear to have the ability to excrete Ni; it has been found in saliva, sweat, tears and milk; biliary excretion may be significant in humans. Hair is also an excretory tissue for Ni.

2.9.1 Biomarkers of exposure

Nickel measurements in the urine, serum or hair may serve as indices of exposure; especially at the end of weekly work period for the biological monitoring for occupational exposure. The normal ranges of Ni concentrations in body fluids or tissues (serum, blood, lung, kidney) are not significantly influenced by age, sex or pregnancy. In the general population, average Ni concentrations in serum and urine are 0.2 and 1–3 µg/L, respectively [104]. In Italy, the reference value for urinary Ni in youngsters from Rome is 0.20–1.23 µg/g creatinine [105] and in a urban population for serum and blood, Ni reference values reported are 0.46 ± 0.36 µg/L and 0.89 ± 0.61 µg/L, respectively [79]. Literature data revealed that Ni concentration in air of 0.1 µg/m^3 is according to a serum Ni level of 0.7 µg/100 mL [106]. There is a lot of data about the assessment of Ni exposure by means of measurements in the urine, faeces, serum, hair, and nasal mucosa [107–109]. Based on biological monitoring data, Sunderman [110] concluded that serum and urine Ni levels were the most useful biomarkers of Ni exposure.

2.9.2 Biomarkers of effect

At the moment, no specific biomarkers of effect for Ni are available, although studies suggested some potential biomarkers. For example, antibodies to hydroxymethyl uracil determined in workers were found significantly increased compared to controls in workers exposed to a mixture of Ni and Cd, as well as in welders [111]. Furthermore, a preliminary study using imaging cytometry of nasal smears obtained from Ni workers indicates that this method may be useful to detect precancerous and cancerous lesions [112].

2.9.3 Biomarkers of susceptibility

A relationship between HLA and Ni sensitivity was observed in individuals who had a contact allergy and positive results in a patch test for nickel [113].

The Ni-sensitive group had a significant elevation in HLA-DRw6 antigen, compared to controls with no history of contact dermatitis. The presence of DRw6 may be monitored to determine the potential risk of individuals to become sensitized to this metal. But, as in the case of biomarkers of effect, no specific biomarkers related to human susceptibility can be found at the moment.

2.10   Thallium

Thallium (Tl) exists in two chemical states; the thallous state is the more common and stable form and thallous compounds are the environmental compounds at which you can be exposed. Thallium is used mostly in the manufacture of electronic devices, switches and closures. Until 1972 this element was used as a rat poison, but was then banned due to its harmful effect to humans. General population can be exposed to Tl by different sources as air, water and food. However, the Tl levels in air and water are very low. The greatest exposure is by contaminated food, mostly homegrown fruits and green vegetables. Cigarette smoking is another Tl source. It has been estimated that the average dietary daily intake of Tl is 2 µg/kg [114]. Occupational exposure is related to that people who work in power plants, cement factories and smelters, where they can get in touch with high concentration of Tl by inhalation and dermal contact. The target organs for absorbed Tl are kidneys and liver. The most common way of excretion is by urine and by faeces in a lesser extent. It is possible to detect urinary Tl (U-Tl) also after 1 h from the exposure. While after 24 h it can be detected also in faeces. Depending upon the tissue, Tl has a biological half-life of 8–10 days. Some effects of Tl poisonings are loss of hair, vomiting, diarrhoea and damage to peripheral nerves. Furthermore, if large amounts of the element are eaten or drunk for short periods of time can cause adverse effects at level of lung, heart, liver and kidney. Death can occur with 1 g of Tl. No information was found on health effects in humans after exposure to smaller amounts of Tl for longer periods. Tl is a suspected human carcinogen [114].

2.10.1   Biomarkers of exposure

The determination of U-Tl is the most widely used biological marker of exposure. Typical Tl levels in non-exposed population are below 1 µg/g creatinine [115]. Urinary levels in cement workers ranged between <0.3 and 6.3 µg/g creatinine [115]. A mean U-Tl level of 76 µg/L was reported in a population living in the vicinity of a cement production plant [116]. Apostoli *et al.* [117] reported mean U-Tl levels of 0.8 and 0.33 µg/L in two groups of workers employed in two cement production plants and two cast iron foundries. An occupational threshold limit of 50 µg/L in urine has been proposed, but should be changed on the basis of new results. Thallium can be detected in blood, but it is cleared from the blood very quickly. Anyway reference values for serum and blood Tl are reported for an urban Italian population: 0.04 ± 0.02 and 0.07 ± 0.04, respectively [78].

2.10.2   Biomarkers of effect

Neurological damage is the primary toxic effect associated with Tl exposure. Electromyographic measurements of nerve conduction velocity and amplitude can be monitored to detect early signs of neurotoxicity. However, since neurological damage occurs with other compounds, these tests are not specific for Tl exposure. Also, the element accumulates in hair. Dark pigmentation of the hair roots and hair loss are common diagnostic features [118]. Depletion and inhibition of several enzymes in the brain (depletion of succinic dehydrogenase and guanine deaminase) have been associated with Tl exposure.

2.10.3   Biomarkers of susceptibility

No biomarkers of susceptibility are at the moment suggested for Tl exposure.

3   Polychlorinated biphenyls (PCBs)

Polychlorinated biphenyls (PCBs) is a group (more than 200, among congeners and isomers) of widespread persistent environmental contaminants, banned by EPA in 1979 [119], with a wide impact on human health, specifically on the endocrine system [120]. PCBs can persist in the environment for very long time and they can also travel long distances in the air far away from where they have been released. In water, a small amount of PCBs may remain dissolved, but most are stick to organic particles. PCBs also bind strongly to soil. In water, small organisms and fish are able to uptake PCBs and they can bioaccumulate PCBs becoming available for animals that feed these organisms as food. General population can be exposed to PCBs by several sources such as using electrical devices and appliances made 30 or more years ago, by eating fish (especially fish caught in contaminated lakes or rivers), meat and dairy products. Exposure by inhalation can occur by breathing air near hazardous waste sites and by drinking contaminated water. Several studies on people exposed to PCBs by contaminated rice oil in Japan [121] and Taiwan [122–124], by contaminated fish and other food products of animal origin and via general environmental exposure are reported; the effects on human of these exposures were widely investigated [125]. Occupational exposure can occur during repair and maintenance of PCBs transformers and other old electrical devices; inhalation is considered a major route of occupational exposure to PCBs and dermal absorption is recognized as important to the accumulation of PCBs in adipose tissue of workers in the capacitor manufacturing industry [126].

PCBs can mimic or block the action of hormones from the thyroid and other endocrine glands [127, 128]. The most commonly observed health effects in humans exposed to large amounts of PCBs are skin complications such as acne and rashes [129]. Furthermore, liver, thyroid, dermal and ocular changes, immunological alterations, neurodevelopment changes, reproductive toxicity and cancer were also observed [125, 130]. Prenatal exposures have been associated with birth weigh reduction [131]. The EPA and the IARC have determined that

12 PCBs are probably carcinogenic to humans (classified in class 2A). Data on the association between occupational exposure to PCBs and a higher cancer risk (especially at digestive system level) were reported [132]. Other papers reported the association between the environmental exposure to PCBs and the breast cancer, even if only in some risk categories [133]. PCBs toxicity depends on multiple variables including the type and structure of PCBs, dose and route of exposure. Research suggests that mechanisms of PCBs toxicity are multifactorial, but appear to involve both the aryl hydrocarbon (Ah) receptor dependent and Ah receptor independent mechanisms [134]. When PCBs (dioxin-like compounds) bind to the Ah receptor, they can disrupt the modulation of normal gene expression determining toxic effects on immunological system [135]. PCBs have been found to induce hepatic Phase I (Cytochrome P450 enzymes) and Phase II enzymes (i.e. uridine diphosphate, glucuronyl transferase and glutathione transferase) [136]. Furthermore, PCBs have shown toxicity in multiple organs and organ systems of the body. At the beginning, PCBs uptake involves liver and muscle due to the high blood perfusion in the liver and the relatively large amount (volume) of muscle. Subsequently, the distribution of PCBs in humans depends on the characteristics of the various congeners. The absorbed PCBs can either be excreted or retained in adipose tissue, skin and specific tissue. Typically, less chlorinated congeners are readily metabolized and excreted. Conversely, highly chlorinated congeners are metabolized very slowly and tend to store in adipose tissue [125]. Due to its high fat content, human milk can also accumulate large quantities of PCBs [137]. Additionally, biotransformation of PCBs can yield hydroxylated metabolites that are retained in the blood. The OSHA's blood limit in 2000 was 25 µg/dL [138].

3.1   Biomarkers of exposure

Since PCBs are lipophilic, they are generally stored in adipose tissue. However, they are also present in serum, blood plasma and human milk [139], especially congeners with a high degree of chlorination and congeners that lack unsubstituted *m*- and *p*- positions that are better candidates for bioaccumulation. Levels of PCBs in blood, body fat and breast milk can be determined but these are not routinely conducted. Both serum and plasma PCBs levels can be significantly affected by serum lipid content (total cholesterol, free cholesterol, triglycerides and phospholipids) because of partitioning between adipose tissue and serum lipids. Therefore, serum or plasma lipid PCBs levels are a better biomarker for body burden than serum or plasma PCBs levels without lipid content corrections [140]. Due to bioaccumulation in adipose tissue, PCBs levels can be measured in tissue samples [141]. PCBs levels in adipose tissue will normally display different compositions when compared to the original PCBs source. Therefore, body burdens of PCBs should be based on the levels of individual congeners or groups of congeners, not on the commercial PCBs mixtures. Accumulation of PCBs also occurs in human milk [142]. High-resolution analytical techniques such as gas chromatography can be used to compare the congeneric composition of PCBs in human milk with commercial mixtures. Chloroacne and other dermal conditions are relevant biomarkers of exposure to PCBs [143, 144]. These conditions are typically found in subjects occupationally exposed. Chloroacne appears when serum PCBs levels are 10 to 20 times greater than the levels found in the general population. Therefore, chloroacne is not a sensitive (or specific) biomarker of PCBs exposure.

3.2 Biomarkers of effect

The Caffeine Breast Test has been used to make known the effects of PCBs. ^{13}C-methyl caffeine is ingested and the activity of the hepatic enzyme cytochrome P4501A2-dependent caffeine 3-N-demethylase is monitored by measuring the quantity of caffeine exhaled as radiolabeled CO_2. This biomarker is not specific, because it is also responsive to polychlorinated p-dibenzo dioxins (PCDDs) and polychlorinated p-dibenzo furans (PCDFs) that also induce the cytochrome P4501A [145]. Evidence in animals suggests that liver enzyme induction is also a sensitive biomarker of PCBs effects, but, again, it is not specific [146]. Mixed-Function Oxidase (MFO) induction has been demonstrated indirectly in PCBs-exposed workers by increased metabolic clearance of antipyrine [147]. A specific human hepatoma cell line, HepG2, has been used to determine dose–response characteristics of a mixture of related PCBs (Aroclor) [148].

3.3 Biomarkers of susceptibility

No biomarkers of susceptibility has been set till now, but it has been observed that people with Gilbert's syndrome are more susceptible to PCBs exposure. This syndrome is a congenital liver disorder consisting in an increasing serum bilirubin level and people who suffered from this syndrome show a decreased capacity to detoxify and excrete PCBs [149].

4 Solvents

Solvents are among the most commonly used chemicals in workplaces. Most solvents used in industry are "organic", petroleum-based chemicals which have powerful properties to dissolve solids. They are often mixtures of several substances and can be extremely hazardous. Some commonly used solvents are acetone, benzene, xylene, hexane, trichloro-ethylene, toluene, methylene-chloride, methyl-ethyl-ketone, perchloro-ethylene, etc. All the above-cited compounds are able to enter in the human body through different ways (inhalation, ingestion and skin absorption). Different solvents have different health effects, which will depend also on how exposure happens, how much and for how long. Short-term exposure can cause dermatitis or skin problems (drying, cracking, reddening or blistering of the affected area) [150] headaches, poor coordination and nausea. In cases of exposure to very high concentrations of solvent vapour, unconsciousness and even death can occur [151]. On the other hand, repeated (long-term) exposure to solvents may affect the brain and the nervous system, liver, blood-forming system, kidneys, fertility of both men and women, foetus in a pregnant woman [152–156]. Some solvents, for example, benzene, can cause cancer at kidney and bladder level [157, 158]. They can also have synergistic effects with other hazards and drugs. This means that the solvent will have greater health effects when it is in combination with other hazardous substances. For all these reasons biomonitoring solvents is a very important goal to avoid human risks. In the next sections are reported reference values and the more useful biomarkers for biomonitoring two solvents widely present and whose health effects are well known: benzene and toluene.

4.1   Benzene

Benzene can be considered an ubiquitous xenobiotic due to its presence at 1% concentration in petrol and in fuel gas. Moreover, cigarettes smoke is a remarkable benzene *indoor* source; in fact, a cigarette burning discharges up to 50 μg of benzene. Also wood fires and eruptions are supplementary firm sources of environmental benzene [159]. Due to the wide presence of benzene in the environment both *indoor* and *outdoor* and the consequent health risk for the general population and workers, the EU set standard limits for exposure to benzene. Benzene is an established human leukemogen, and it may also cause non-Hodgkin's lymphoma and other cancers [159], although its ability to cause cancer at low levels of exposure is unknown. Studies on worker populations showed a range of haematotoxic effects including anaemia, leukopenia and thrombocytopenia [160]. In the past few years, due to its carcinogenic properties, benzene was progressively, but not completely, replaced by other aromatic and aliphatic compounds. Benzene is a lipophilic solvent and its uptake occurs mainly by inhalation, by ingestion or through the skin. Its distribution in the biological matrices depends on the water/air and the tissue/blood distribution coefficients. Benzene is disseminated in several human matrices such as blood, bone marrow, adipose tissue and liver. Moreover, benzene is transported from blood to lungs where it can be excreted. Its half-life is about 9 h, but it can reach the 24 h if it is accumulated in the adipose tissue from which is released slowly [159]. In the human body, its metabolism starts with the activity by cytochrome P4502E1, and perhaps other cytochromes P450, forming at the first step the benzene-epoxide and then the phenol. An alternative mechanism for phenol formation is by the benzene oxide-oxepin system. When benzene oxide is the first product, it can rearrange non-enzymatically to form phenol. Alternatively, benzene oxide can be hydrated via epoxide hydrolase to yield 1,2-benzene dihydrodiol, which can in turn be oxidized via dihydrodiol dehydrogenase to form catechol. The reaction of benzene oxide with glutathione catalysed by GST leads to the formation of the premercapturic acid. Phenol can be further hydroxylated to form hydroquinone or catechol. Catechol can form *trans-trans*-muconic aldehyde, which can be oxidized to form *t,t*-muconic acid. Phenol, catechol and hydroquinone, after conjugation with sulphate or glucuronide are excreted by the urinary way [161]. All these biotransformations occur mainly at hepatic level. Benzene and its metabolites can also give rise at metabolic bioproducts and adducts binding to DNA or proteins, such as albumin and haemoglobin [162]. Laboratory studies on mice demonstrated that the main way of excretion of the benzene's metabolites is the urinary way. For humans, these data are not disproved and moreover it has been shown that just a small amount of benzene is excreted by pulmonary way and less than 1% of the absorbed dose is excreted like benzene unmodified in urine [159]. On the other hand, phenol's urinary excretion reaches the 74–87% of the adsorbed dose after a few hours from the exposure as far as return at the basal excretion after 16 h [163].

4.1.1 Biomarkers of exposure

A few biological markers of exposure are available for benzene, including expired benzene, blood benzene and urinary metabolites of each pathway [163]. Urinary phenol, S-phenylmercapturic acid (S-PMA) and t-,t-muconic acid are usually used as biomarkers of benzene exposure in occupational investigations [163, 164]. The efficacy of urinary phenol as a biomarker of benzene exposure is limited to inhalation exposures at air concentrations exceeding 3 mg/m^3 and has been found to be proportional to benzene concentrations at exposures as high as 620 mg/m^3. Additional sources of urinary phenol due to diets and ingestion of medicine have precluded its use as a biomarker of environmental benzene exposures [164]. Urinary S-PMA it was found to be increased in a dose–response way to benzene in the urine of coke production workers, in a similar manner as urinary phenol, but urinary S-PMA seems to be a more sensitive biomarker of benzene exposure than urinary phenol [165]. Urinary S-PMA appears to be a specific and sensitive biomarker of benzene and it has the potential to be a useful biomarker of low-level benzene exposures. However, it is excreted in very low quantities (μg/g creatinine) and requires a complex analytical method to be quantified. Urinary t,t-muconic acid has been used as a biomarker of sub-mg/m^3 levels of exposure, the upper range of environmental benzene exposure; but other sources of this urinary metabolite have been identified, such as metabolism of sorbic acid. The ACGIH/BEI set value for S-PMA in urine is 25 μg/g creatinine, while the reference value for non-smokers is < 5 μg/g creatinine. The ACGIH/ BEI value for t,t-muconic acid in urine is 500 μg/g creatinine, while the reference value for non-smokers is <5 mg/g creatinine. Blood-benzene is another biomarkers; the results of the NHANES in general population pointed out that blood benzene concentrations of smokers were elevated compared to those of non-smokers. For blood benzene the reference value was <0.5 μg/L. S-phenyl cysteine in blood of workers, detected by an ELISA method, is a new biomarkers of exposure proposed during the last 6th *International Symposium on Biological Monitoring in Occupational and Environmental Health* [166].

4.1.2 Biomarkers of effect

The association between benzene exposure and the appearance of structural and numerical chromosomal aberrations in human lymphocytes suggests that benzene may be considered as a human clastogen. In animals benzene induced cytogenetic effects, including chromosome and chromatid aberrations, SCEs and *micronuclei* [159]. Several lines of evidence also indicate that benzene is - genotoxic in humans under occupational exposure conditions. However, these studies lacked good exposure monitoring data, involved multiple chemical expo-sures and were often poorly designed, with inappropriate control groups. So that *micronuclei*, chromosome and chromatid exchanges, aberrations, etc. cannot be considered validated biomarkers of effects, at the moment.

4.1.3   Biomarkers of susceptibility

Previous studies have indicated that benzene toxicity mainly results from its intermediate reactive metabolites. Hypothesis is that the deficient or altered activity of enzymes involved in benzene metabolism such as CYP2E1 (family of cytochromes, that can oxidize alcohol to form acetaldehyde), myeloperoxidase (MPO), quinone oxidoreductase (NQO1) and GST would significantly affect susceptibility to benzene toxicity [167]. Genetic polymorphisms in genes encoding CYP2E1, MPO, NQO1 and GSTs might be responsible for human susceptibility because they might affect enzyme activity. There are studies that suggest this hypothesis but, at the moment, stronger confirmation of these findings is needed [167].

4.2   Toluene

Due to the presence of toluene in many consumer products, humans can be exposed to toluene *indoor* and *outdoor* by breathing contaminated air deriving from the use of such products and by breathing contaminated workplace air or automobile exhaust. Also cigarette smoke is a slight source of toluene [168]. Low levels of toluene can affect the nervous system determining tiredness, confusion, weakness, nausea, memory, appetite and colour vision loss [169], while high levels can cause unconsciousness and even death [170, 171]. At the moment the IARC and the US Department of Health and Human Services have not classi-fied toluene as human carcinogen, but Burgaz *et al.* [172] reported that people exposed to organic solvents, mainly toluene, working in the shoe manufacture have an increased risk for cancer in particular for nasal cancer and leukaemia. Toluene is quickly absorbed from the respiratory and gastrointestinal tracts but the 75% of the assimilated toluene is generally removed within 12 h by the breathing out or by the urine; toluene is generally biotransformed into less harmful compounds such as hippuric acid. In particular the initial step of the toluene metabolism consists of the side-chain hydroxylation to form benzyl alcohol. This first step is catalysed mainly by the cytochrome P450 (CYP) isozyme, CYP2E1. Then the benzyl alcohol is oxidized to benzoic acid that is conjugated with glycine to form hippuric acid, quickly excreted by urine with a clearance of 700 mL/min. A very small amount (<5%) of absorbed toluene is converted by CYP1A2, CYP2B2 or CYP2E1 to *o*- or *p*-cresol and excreted as sulphate or glucuronate conjugates by the urine [173, 174].

4.2.1 Biomarkers of exposure

Many biomarkers to measure the exposure to toluene, such as its concentration or its breakdown products in exhaled air, urine and blood have been proposed. The most accurate biomarker is the presence of toluene in serum or blood, but measurements of toluene or its metabolites in urine are often preferred due to the less invasive sampling [175]. However, in must be stressed out that the presence of these compounds in the urine is not a definitive proof of the toluene exposure, because they are also produced by normal metabolism from diet [176]. In addition, the background levels of these metabolites may be affected by individual variability [177], ethnic differences [178] or other factors such as alcohol consumption and smoking practice [176, 179]. The S-p-toluyl-mercapturic acid levels in urine may also be a useful biological marker of toluene exposure [180]. Despite these limitations the ACGIH [181], compared to the TLV-TWA limit of 50 ppm, sets up the measurements of o-cresol (at 0.5 mg/L) and hippuric acid (at levels of 1.6 g/g creatinine) in urine and toluene in blood (at levels of 0.05 mg/L) to assess the exposure of workers to toluene in the workplace at the end of a work shift [181].

4.2.2 Biomarkers of effect

There are no specific biomarkers used to characterize the consequences of the toluene exposure. Wiwanitkit *et al.* [182] tried to use SCE levels as a biomarker of effect, calculating the correlation between SCE and urine hippuric acid among Thai traffic police. But they found that high exposure to toluene seems to be not perfectly related to high SCE content. Furthermore, toluene seems not to form adducts and it has a little known health risks at environmental levels. Mendoza-Cantú *et al.* [183] suggested that CYP2E1 mRNA content in peripheral lymphocytes could be a sensitive and non-invasive biomarker for the monitoring of toluene effects in exposed people.

4.2.3 Biomarkers of susceptibility

No susceptibility biomarkers have been established for toluene. Toluene is metabolized by cytochrome P4502E1, and then such polymorphisms may be associated with altered susceptibility. A number of these biomarkers are possible but they need further investigation. Also the polymorphic differences in DNA repair enzymes might also be important predictors of susceptibility [184].

5 Pesticides

The term pesticide includes different kind of compounds such as insecticides, fungicides, herbicides, rodenticides, molluscicides and others. Insecticides cover a huge fraction of the total pesticide usage and the principal classes of compounds that have been used as insecticides are organophosphorus, organochlorine, carbamate and pyrethroid compounds, and various inorganic compounds. Pesticides can cause adverse effects to human, indeed every year three million cases of pesticide poisonings are scheduled, of which 220,000 are fatal [185]. The sources of exposure include water, air and food; the pesticide's uptake occurs mainly through the skin and eyes, by inhalation or by ingestion. Absorption resulting from dermal exposure is the most important route of uptake for exposed workers [186]. Acute toxic effects are easily identified, whereas the effects resulting from long-term exposure to low doses are often difficult to recognize [187]. Workers exposed to organophosphorus pesticides alone or in combination with organochlorine or other pesticides showed changes in their liver enzyme activities [188]. Increased risks for spontaneous abortion and decreased birth weight were reported in a population in Colombia where exposure to many pesticides occurred [189]. Genotoxic effects are considered among the most serious of the possible side effects of agricultural chemicals; effects include heritable genetic diseases, carcinogenesis, reproductive dysfunction and birth defects. The IARC has classified 56 pesticides as carcinogenic after laboratory tests to animals [188]. Increased mortality from lung cancer in farmers from United States [190] and from Florida [191], and an increased risk of lung cancer in Germany [192] have been reported. Among farmers in the Piedmont (Italy) increased risk for skin cancer and malignant lymphomas was also observed [193]. In Nebraska, the risk of non-Hodgkin's lymphoma increased with frequency of use of organophosphorus insecticides among farmers [194]; in a Washington State study, non-Hodgkin's lymphoma was associated with potential contact with chlordane and dichlorodiphenyltrichloroethane [195].

5.1   Biomarkers of exposure

The organophosphorus pesticides are rapidly hydrolysed to six dialkylphosphate metabolites which are detectable in urine and may be measured for several days after exposure. The maximum concentration of urinary dialkylphosphates was reached after 17–24 h, and the half-life was 30 h. The analysis of six dialkylphosphates metabolites in urine, namely, dimethylphosphate, dimethylthiophosphate, dimethyldithiophosphate, diethyl-phosphate, diethyl-thiophosphate and diethyldithiophosphate is a useful approach to biomonitoring for organophosphorus pesticides [196]. These metabolites may be measured for several days after exposure. The maximum concentration of urinary dialkylphosphates was reached after 17–24 h, and the half-life was 30 h. Dimethylthiophosphate has been found to be the dominant of the three dimethyl metabolites in the 1999–2000 NHANES population survey [197]. The urine determination of these metabolites is used to monitor occupational exposure to organophosphorus pesticides, because the ease of the sample collection and of the analytical procedure. Main disadvantage is that urine output varies, and therefore the concentration of organophosphorus pesticides may vary. This may be solved by creatinine correction in urine samples. Moreover, these urinary metabolites are not pesticide specific, and they may enter the body from other exposure sources. Despite these inconveniences, measurements of dialkylphosphates metabolites are one of the commonly used markers of organophosphorus pesticides exposure.

Blood-hexachlorobenzene (HCB) is a biomarker of exposure to HCB [198], a fungicide largely used in agriculture [199]. General exposure is from food (olive oil) and contaminated water; absorptions are through respiratory system and skin. The only known HCB metabolite in human is the pentachlorophenol. Some studies report that HCB can be stored in the adipose tissues. Long-term exposure can lead to central nervous system alterations and porphyria; HCB can also induct microsomial hepatic enzymes [200].

Blood-Lindane is biomarker of exposure to Lindane (γ-hexachlorocyclo-hexane), a chlorinated hydrocarbon determined by the EPA as carcinogenic, although the evidences are not sufficiently clear. In human, Lindane is quickly metabolized and its main urinary metabolites are 2,3,5-; 2,4,5-; and 2,4,6-trichlorophenol. Neurological diseases are associated with a blood-Lindane concentration of 0.02 mg/L. Biological limit values of 0.02 mg/L of Lindane in the whole blood and of 0.025 mg/L in serum (or plasma) have been established in Germany. The World Health Organization (WHO) recommends a value of 0.02 mg/L in the whole blood [201]. Urinary p-nitrophenol (PNP) is frequently used as a marker of exposure to parathion and methyl parathion [202, 203]. The EPA and IARC have determined that methyl parathion is not classifiable as human carcinogen. After the exposure to parathion and methyl parathion, acetylcholine at level of the synaptic junctions is accumulated. The PNP is implicated in the detoxification metabolism of the two toxic compounds during Phase I [204], and it is eliminated through the urine either as the free phenol or as its glucuronide or sulphate ester [205]. After an acute exposure, PNP is excreted in urine within a few days, while chronic exposure to methyl parathion, as with any non-persistent organic toxicant, can result in steady-state excretion of the metabolite. The U-PNP DFG/BAT value is 500 µg/L, while the reference is <0.01–0.03 mg/L.

Among the metabolites produced by ethylene-bisdithiocarbamates, ethylene-thiourea (ETU) is one of them. Additionally, ETU can occur as an impurity in commercial ethylene-bisdithiocarbamates formulations [206]). The ETU antithyroid activity and its teratogenic properties are well known; ETU was classified as possible carcinogenic to humans by IARC (group 2B) [207], but recently was re-classified in the group 3, i.e. compound with inadequate evidence of carcinogenicity [208]. The urinary excretion of ETU reaches the percentage of 52 and 86% of the total dose after 24 and 48 h, respectively. Due to the fact that ethylene-thiourea has been detected also in cigarette smoke and in wine [209], when it is measured as a metabolite of ethylene-bisdithiocarbamates information on life habits has to be considered. Reference value for ETU is 2 µg/g creatinine.

5.2   Biomarkers of effect

The reduction in cholinesterase activity is a biochemical biomarker of organophosphates exposure widely used, which may involve destructive sampling (brain acetylcholinesterase (AchE)) or non-destructive sampling (serum butyryl-cholinesterase). Diagnostic kits to measure specific activities of blood esterases would be of considerable help in measuring exposure to organophosphorus insecticides in the field [210]. Measurement of AChE in red blood cells is applied to assess cholinergic effects [211, 212].

5.3 Biomarkers of susceptibility

Many enzymatic isoforms have been suggested to contribute to the individual cancer susceptibility [213, 214]. Cytochrome P450 2E1 (CYP2E1), glutathione *S*-transferase M1, glutathione *S*-transferase-*θ*-1 (GST-T1), *N*-acetyltransferase 2 (NAT2) and *para*oxonase 1 (PON1), are the polymorphic genes related to the metabolism of these xenobiotics. In particular, CYP2E1 is a cytochrome P450 superfamily involved in the metabolism of many indirect carcinogens [215], pesticides included. The cancer risk can be, then, modified through the alteration of CYP2E1 enzyme activity. Significant associations between CYP2E1 polymorphism and cancer risk was, in fact, reported in a number of studies [216]. The GSTs family is responsible for the glutathione conjugation of various reactive species of many chemicals including pesticides [217]. PONs are responsible for metabolism of organophosphate insecticides [218, 219], in particular serum PON1 activity plays a major role in the metabolism of organophosphates (e.g. Chlorpyrifos oxon, Diazoxon which are the active metabolites of Chlorpyrifos and Diazinon). PON1 is responsible for the hydrolysis and deactivation of organophosphates and is expressed in liver and in blood, where it is tightly bound to high-density lipoproteins. The role of PON1 in determining susceptibility to organophosphates insecticides has been studied in transgenic animal models. PON1 knockout mice are exquisitely sensitive to the acute toxicity of Chlorpyrifos oxon and Diazoxon. At this time, however, biomarkers of susceptibility have not been adequately studied and validated to incorporate into clinical and surveillance practice.

6 Aromatic amines

Aromatic amines are often classified as suspected or noted human cancerogenous compounds. Clayson and Garner [220] suggested that to be carcinogenic, an aromatic amine had to possess two or three conjugated aromatic ring systems, have the amino group substituted in the aromatic ring in the *para* position to a conjugated aromatic system and groups such as methyl, methoxil or fluorine substituted in specific positions relative to the amino group. Aromatic amines have been assayed for carcinogenicity. The study of Case *et al.* [221] in Great Britain provided the first systematic evidence of the carcinogenicity of specific aromatic amines. The IARC since from 1972 stated the implication of 2-naphthylamine, benzidine and 4-aminobiphenyl in bladder cancer in workers exposed to these compounds. At the moment, only 2-naphthylamine, benzidine and 4-aminobiphenyl are classified by IARC in Group 1 (human cancerogen). On the other hand, others compounds are classified like probable cancerogens

(Group 2, for instance, 4-chloroaniline, 3,3'-dichlorobenzidine, *o*-toluidine), and others like not cancerogenous classifiable (Group 3, such as 2,4,5-trimethylaniline and aniline) [222]. Occupational exposure by inhalation and ingestion is the main way of exposure to aromatic amines. Non-occupational exposures are through food, water, soil, medicines and make-up. Furthermore, tobacco smoke is an important source of aromatic amines, such as aniline, *o*-toluidine, *m*-toluidine, *p*-toluidine, 2,3-dimethylaniline, 2,4-dimethylaniline, 2,5-dimethylaniline, 2,6-dimethylaniline, 2-naphthylamine and 4-aminobiphenyl. In most *indoor* environments with no contamination from cigarette smoke the aromatic amines levels can be below 20 ng/m^3, whereas in the presence of smokers higher values were observed [223]. In the next sections are reported some biomarkers and reference values for biomonitoring aniline and orto-toluidine, two of the major aromatic amines known.

6.1   Aniline

The general population may be exposed to aniline by eating food, by drinking water containing aniline, through drugs assumption (cardiovascular drugs containing nitro-glycerine) and tobacco smoke – a cigarette contains 50.6–577 ng of aniline (1–10 µg per packet). That results in a urinary aniline excretion by a moderate smoker between 0.4 and 4.0 µg/L [224]. But exposure for the general population is quite moderate; on the other hand, people working in dyes, varnishes, herbicides and explosives industries may be heavily exposed to aniline. Pesticides degradation is another source of aniline. Aniline can damage haemoglobin, carrying at the consequent methaemoglobinaemia and cyanosis. Other symptoms related to aniline exposure are headaches and dizziness, altered heartbeat, convulsions and even coma and death. Long-term exposure to aniline cause symptoms similar to those occurred in acute exposure. Regards the aniline carcinogenicity, the IARC did not classified aniline as human carcinogen, while EPA determined that aniline is a probable human carcinogen.

6.1.1   Biomarkers of exposure
Aniline can be measured in the urine. Urinary aniline (U-aniline) is a biomarker of aniline exposure, but it does not show how much or how recently the exposure has been done. A breakdown product of aniline in the body produced by the cytochrome P450 family is the *p*-aminophenol, that can also be measured as sulphates or glucuronates in the urine up to 24 h after exposure. However, this breakdown product is not specific for aniline exposure. The ACGIH/BEI is 50 mg/g creatinine [225]. Methaemoglobin levels in blood can be measured aniline exposure, but exposure to many other chemicals also increase methaemoglobin levels. However, the ACGIH/BEI for blood methaemoglobin is 1.5% of haemoglobin, while the reference value is <1%.

6.1.2   Biomarkers of effect
Haemoglobin adducts (Hb-A) have been used to establish a biological tolerance value for occupational exposures (BAT value) for the first time with aniline (100 µg of released aniline/L of blood). This value is considered to correspond with

a maximum methaemoglobin level of 5% or an aniline concentration in urine of 1 mg/L. However, average occupational exposure and the total burden are better reflected by adduct levels than by the concentration of aniline in urine [226].

6.1.3 Biomarkers of susceptibility

Recently, experimental evidence revealed that the heterocyclic aromatic amines may be involved as causal factors in the human breast cancer [227]. In humans these amines require an enzymatic activation to promote carcinogenesis after binding to DNA. N-acetyltransferase 2 (NAT2) activity is very important due to this activation. The relationship between the NAT2 genetic polymorphisms and breast cancer seems to be checked. So, the NAT2 polymorphisms could be used as biomarkers of susceptibility [228].

6.2 ortho-Toluidine

General population may be exposed to low concentrations of o-toluidine in air, food or by dermal contact with commercial products. Tobacco smoke is an extra source of o-toluidine in the environment; o-toluidine amount in cigarette smoke is approximately 32 ng/cigarette [229]. Occupational exposure to higher concentrations of o-toluidine occurs by inhalation and skin contact. The greatest exposures are for dye makers and pigment makers [230]. The National Institute for Occupational Safety and Health (NIOSH) indicated that from 1981 to 1983 approximately 30,000 workers were exposed to o-toluidine [231]. o-Toluidine has the potential to produce skin and eye irritation; headaches, cyanosis, dizziness, drowsiness and haematuria are other symptoms. The principal signs of toxicity after an acute and short-term exposure are methaemoglobinaemia and related effects in the spleen. The toxicological consequences deriving from a long-term exposure to o-toluidine are known. For example, some epidemiological studies suggested that occupational exposure to o-toluidine is associated with an increased risk of bladder cancer [232–234]. o-toluidine is classified by the IARC as probably carcinogenic for humans, while is considered to be a proven human carcinogen by the German Commission for the Investigation of Health Hazards of Chemical Compounds in the Work Area (MAK Commission) [235, 236]. In standard bacterial mutagenicity tests o-toluidine resulted not mutagenic, but resulted clastogenic in mammalian cells in vitro.

6.2.1 Biomarkers of exposure

The analytical monitoring of urine for o-toluidine and its N-acetyl metabolites is a useful tool of assessing occupational exposure. Blood is generally another biomarker of exposure for o-toluidine.

6.2.2 Biomarkers of effect

Haemoglobin adducts (Hb-A) are considered to be excellent surrogate markers of aromatic amine bioactivation to their ultimate carcinogenic metabolites [237, 238] as demonstrated by several studies [239].

6.2.3 Biomarkers of susceptibility

Data on the implication of cytochrome P450 (CYP) 1A1 and 1A2 in metabolism of *o*-toluidine in rats have been reported [240]. Smokers resulted in a higher CYP1A2 activity compared to non-smokers [241]. So they could be used as biomarkers of susceptibility after further studies.

References

[1] Center for Disease Control and Prevention (2006) Third National Report on Human Exposure to Environmental Chemicals, URL: http://www.cdc.gov/exposurereport.

[2] European Environment and Health Committee, URL: http://www.euro.who.int/eehc.

[3] National Research Council (NRC) (1989) Biologic Markers in Reproductive Toxicology. National Academy Press, Washington, DC, p. 420.

[4] McCarthy F., Shugart L.R. (1990) Biomarkers of environmental contamination. Lewis Publications, Chelsea, USA, p. 472.

[5] Depledge M.H., Fossi M.C. (1994) The role of biomarkers in environmental assessment. Invertebrates. Ecotoxicology, 3, 161–72.

[6] Alessio L., Berlin A., Roi R. (1983) Human biological monitoring of industrial chemicals series. Luxembourg, Office for Official Publications of the Commission of European Communities (Industrial Health and Safety Series).

[7] Alessio L., Berlin A., Roi R. (1984) Biological indicators for the assessment of human exposure to industrial chemicals. Luxembourg, Office for Official Publications of the Commission of the European Communities (Industrial Health and Safety Series - EUR 8903 EN).

[8] Alessio L., Berlin A., Roi R. (1986) Biological indicators for the assessment of human exposure to industrial chemicals. Luxembourg, Office for Official Publications of the Commission of the European Communities (Industrial Health and Safety Series - EUR 10704 EN).

[9] Alessio L., Berlin A., Roi R. (1987) Biological indicators for the assessment of human exposure to industrial chemicals. Luxembourg, Office for Official Publications of the Commission of the European Communities (Industrial Health and Safety Series - EUR 11135 EN).

[10] Alessio L., Berlin A., Roi. R. (1988) Biological indicators for the assessment of human exposure to industrial chemicals. Luxembourg, Office for Official Publications of the Commission of the European Communities (Industrial Health and Safety Series - EUR 11478 EN).

[11] Alessio L., Berlin A., Roi R. (1989) Biological indicators for the assessment of human exposure to industrial chemicals. Luxembourg, Office for Official Publications of the Commission of the European Communities (Industrial Health and Safety Series - EUR 12174 EN).

[12] UK HSE (1991) Guidance on Laboratory Techniques in Occupational Medicine, 5th ed., Health and Safety Executive, Library and Information Services, London, p. 117.

[13] American Conference of Governmental Industrial Hygienists (ACGIH) (1992) 1992–1993 Threshold limit values for chemical substances and physical agents and biological exposure indices. Cincinnati, Ohio, American Conference of Governmental Industrial Hygienists.

[14] Deutsche Forschungsgemeinschaft (DFG) (1992) Commission for the Investigation of Health Hazards of Chemical Compounds in the Work Area: MAK- and BAT-values. VCH, Weinheim.

[15] Bond J.A., Wallace L.A., Osterman-Golkar S., Lucier G.W., Buckpitt A., Henderson R.F. (1992) Assessment of exposure to pulmonary toxicants: use of biological markers. Fundamental and Applied Toxicology, 18, 161–174.

[16] Grandjean P. (1995) Biomarkers in epidemiology. Clinical Chemistry, 41, 1800–1803.

[17] Costa L., Richter R., Li W., Cole T., Guizzetti M., Furlong C. (2003) Paraoxonase (PON1) as a biomarker of susceptibility for organophosphate toxicity. Biomarkers, 8(1), 1–12.

[18] National Library of Medicine (2000) Aluminum Fluoride, Hazardous Substances Data Base. National Library of Medicine, Washington DC, USA.

[19] Lui J.Y., Stemmer K.L. (1990) Interaction between aluminum and zinc or copper and its effects on the pituitary-testicular axis. II. Testicular enzyme and serum gonadotropin assay. Biomedical and Environmental Sciences, 3(1), 11–19.

[20] Burge P.S., Scott J.A., McCoach J. (2000) Occupational asthma caused by aluminium. Allergy, 55(8), 779–780.

[21] Abbate C., Giorgianni C., Brecciaroli R. (2003) Spirometric function in non-smoking workers exposed to aluminum. American Journal of Industrial Medicine, 44(4), 400–404.

[22] Korogiannos C., Babatsikou F., Tzimas S. (1998) Aluminum compounds and occupational lung disease. European Respiratory Journal, 12(Suppl. 28), 139S.

[23] van der Voet G.B., Schijns O., de Wolff F.A. (1999) Fluoride enhances the effect of aluminum chloride on interconnections between aggregates of hippocampal neurons. Archives of Physiology and Biochemistry, 101(1), 15–21.

[24] Suarez-Fernandez M.B., Soldado A.B., Sanz-Medel A., Vaga J.A., Novelli A., Fernandez-Sanchez M.T. (1999) Aluminum-induced degeneration of astrocytes occurs via apoptosis and results in neuronal death. Brain Research, 835(2), 125–136.

[25] Kausz A.T., Antonsen J.E., Hercz G., Pei Y., Weiss N.S., Emerson S., Sherrard D.J. (1999) Screening plasma aluminum levels in relation to aluminum bone disease among asymptomatic dialysis patients. American Journal of Kidney Disease, 34(4), 688–693.

[26] Altmann P., Cunningham J., Dhanasha U., Ballard M., Thompson J. (1999) Disturbance of cerebral function in people exposed to drinking water contaminated with aluminium sulphate: retrospective study of the Camelford water incidence. British Medical Journal, 319, 807–811.

[27] D'Haese P.C.D., Couttenye M.M., Goodman W.G. (1995) Use of the low-dose desferrioxamine test to diagnose and differentiate between patients with aluminum-related bone disease, increased risk for aluminum toxicity, or aluminum overload. Nephrology Dialysis Transplantation, 10, 1874–1884.

[28] Huang J.Y., Wu M.S., Wu C.H. (2001) The effect of an iron supplement on serum aluminum level and desferrioxamine mobilization test in hemodialysis patients. Renal Failure, 23(6), 789–795.

[29] Sakurai T., Kojima C., Kobayashi Y., Hirano S., Sakurai M.H., Waalkes M.P., Himeno S. (2006) Toxicity of a trivalent organic arsenic compound, dimethylarsinous glutathione in a rat liver cell line (TRL 1215). British Journal of Pharmacology, 149, 888–897.

[30] Holson J.F., Stump D.G., Clevidence K.J., Knapp J.F., Farr C.H. (2000) Evaluation of the prenatal developmental toxicity of orally administered arsenic trioxide in rats. Food and Chemical Toxicology, 38(5), 459–466.

[31] Marcus W.L., Rispin A.S. (1988) Threshold carcinogenicity using arsenic as an example, in: Cothern C.R., Mehlman M.A., Marcus W.L. (Eds.), Advances in Modern Environmental Toxicology, Vol. XV. Risk Assessment and Risk Management of Industrial and Environmental Chemicals. Princeton Scientific Publishing Company, Princeton, NJ, USA, pp. 133–158.

[32] Goyer R.A. (1991) Toxic effects of metals, in: Amdur M., Doull J., Klassen C., (Eds.), Casarett and Doull's Toxicology. The Basic Science of Poisons. Pergamon Press, New York, NY, USA, pp. 623–680.

[33] Shi H., Hudson L.G., Ding W., Wang S., Cooper K.L., Liu S. (2004) Arsenite causes DNA damage in keratinocytes via generation of hydroxyl radicals. Chemical Research in Toxicology, 17, 871–878.

[34] Vahter M. (1986) Environmental and Occupational exposure to inorganic Arsenic. Acta Pharmacologica et Toxicologica, 59, 31–34.

[35] Franzblau A., Lilis R. (1989) Acute arsenic intoxication from environmental arsenic exposure. Archives of Environmental Health, 44, 385–390.

[36] Karagas M.R., Tosteson T.D., Blum J., Morris J.S., Baron J.A., Klaue B. (1998) Design of an epidemiologic study of drinking water arsenic exposure and skin and bladder cancer risk in a U.S. population. Environmental Health Perspectives, 106(Suppl. 4), 1047–1050.

[37] Hindmarsh J.T., McCurdy R.F. (1986) Clinical and environmental aspects of arsenic toxicity. Critical Reviews in Clinical Laboratory Sciences, 23(4), 315–347.

[38] Goebel H.H., Schmidt P.F., Bohl J., Tettenborn B., Krämer G., Gutmann L. (1990) Polyneuropathy due to acute arsenic intoxication: biopsy studies. Journal of Neuropathology and Experimental Neurology, 49(2), 137–149.

[39] Woods J.S., Southern M.R. (1989) Studies on the etiology of trace metal-induced porphyria: effects of porphyrinogenic metals on coproporphyrinogen oxidase in rat liver and kidney. Toxicology Applied Pharmacology, 97, 183–190.

[40] Sardana M.K., Drummond G.S., Sass S. (1981) The potent heme oxygenase inducing action of arsenic in parasiticidal arsenicals. Pharmacology, 23, 247–253.

[41] Menzel D.B., Rasmussen R.E., Lee E., Meacher D.M., Said B., Hamadeh H., Vargas M., Greene H., Roth R.N. (1998) Human lymphocyte heme oxygenase 1 as a response biomarker to inorganic arsenic. Biochemical and Biophysical Research Communications, 250(3), 653–656.

[42] Zorn H.R., Stiefel T.W., Beuers J., Schlegelmilch R. (1988) Beryllium, in: Seiler H.G., Sigel H. (Eds.), Handbook on Toxicity of Inorganic Compounds. Marcel Dekker, Inc., New York, NY, USA, pp. 105–114.

[43] Williams W.J., Williams W.P. (1983) Value of beryllium lymphocyte transformation tests in chronic beryllium disease and in potentially exposed workers. Thorax, 38, 41–44.

[44] Saltini C., Winestock K., Kirby M., Pinkston P., Crystal R.G. (1989) Maintenance of alveolitis in patients with chronic beryllium disease by beryllium-specific helper T cells. New English Journal of Medicine, 320, 1103–1109.

[45] Maier L.A., Sawyer R.T., Tinkle S.S., Barker E.A., Balkissoon R.C., Newman L.S. (2001) IL-4 fails to regulate in vitro beryllium-induced cytokines in berylliosis. European Respiratory Journal, 17, 403–415.

[46] Finch G.L., Mewhinney J.A., Hoover M.D. (1990) Clearance, translocation, and excretion of beryllium following acute inhalation of beryllium oxide by beagle dogs. Fundamental and Applied Toxicology, 15, 231–241.

[47] Furchner J.E., Richmond C.R., London J.E. (1973) Comparative metabolism of radionuclides in mammals: VII. Retention of beryllium in the mouse, rat, monkey and dog. Health Physics, 24, 292–300.

[48] Morgareidge K., Cox G.E., Bailey D.E., Gallo M.A. (1977) Chronic oral toxicity of beryllium in the rat. Toxicology and Applied Pharmacology, 41, 204–205.

[49] Stokinger H.E., Sprague G.F., Hall R.H. (1950) Acute inhalation toxicity of beryllium. I. Four definitive studies of beryllium sulfate at exposure concentrations of 100, 50, 10 and 1 mg per cubic meter. Archives of Industrial Hygiene and Occupational Medicine, 1, 379–397.

[50] Krachler M., Rossipal E., Micetic-Turk D. (1999) Trace of element transfer from the mother to the newborn – investigations on triplets of colostrum, maternal and umbilical cord sera. European Journal of Clinical Nutrition, 53(6), 484–494.

[51] Paschal D.C., Ting B.G., Morrow J.C. (1998) Trace mescals in urine of United States residents: reference range concentrations. Environmental Research, 76(1), 53–59.

[52] Kanarek D.J., Wainer R.A., Chamberlin R.I., (1973) Respiratory illness in a population exposed to beryllium. American Review of Respiratory Disease, 108, 1295–1302.

[53] Reeves A.L. (1986) Beryllium, in: Friberg L., Nordberg G.F., Vouk V.B. (Eds.), Handbook on the Toxicology of Metals, 2nd edn. Vol. II: Specific metals. Elsevier Science Publishers, New York, pp. 95–116.

[54] Tsalev D.L., Zaprianov Z.K. (1984) Atomic Absorption Spectrometry in Occupational and Environmental Health Practice. CRC Press, Boca Raton, FL, p. 252.

[55] Fontenot A.P., Torres M., Marshall W.H. (2000) Beryllium presentation to CD4+ T cells underlies disease-susceptibility HLA-DP alleles in chronic beryllium disease. Proceedings of the National Academy of Sciences, 97(23), 12717–12722.

[56] Lombardi G., Germain C., Urenk J. (2001) HLA-DP allele-specific T cell responses to beryllium account for DP-associated susceptibility to chronic beryllium disease. Journal of Immunology, 166, 3549–3555.

[57] Richeldi L., Sorrentino R., Saltini C. (1993) HLA-DPB1 glutamate 69: a genetic marker of beryllium disease. Science, 262(5131), 242–244.

[58] Richeldi L., Kreiss K., Mroz M.M. (1997) Interaction of genetic and exposure factors in the prevalence of berylliosis. American Journal of Industrial Medicine, 32(4), 337–340.

[59] Agency for Toxic Substances and Disease Registry (ATSDR) (1999) Toxicological Profile for Cadmium (Final Report). NTIS Accession No. PB99-166621. Atlanta, GA, p. 434.

[60] International Agency for Research on Cancer (IARC) (1993) Group 1 Cadmium and cadmium compounds. URL: http://www.inchem.org/documents/iarc/vol58/mono58-2.html.

[61] Lauwerys R.R., Bernard A.M., Roels H.A. (1994) Cadmium: exposure markers as predictors of nephrotoxic effects. Clinical Chemistry, 40(7), 1391–1394.

[62] Elinder C.G. (1985) Normal values for cadmium in human tissue, blood and urine in different countries, in: Friberg L., Elinder C.G., Kjellstrom T., et al. (Eds.), Cadmium and Health: A Toxicological and Epidemiological Appraisal. Vol. I. Exposure, Dose, and Metabolism. Effects and Response. CRC Press, Boca Raton, FL, pp. 81–102.

[63] Roels H.A., Lauwerys R.R., Buchet J.P., Bernard A., Chettle D.R., Harvey T.C., Al-Haddad I.K. (1981) In vivo measurement of liver and kidney cadmium in workers exposed to this metal: its significance with respect to cadmium in blood and urine. Environmental Research, 26(1), 217–240.

[64] World Health Organization (WHO) (1980) Recommended health-based limits in occupational exposure to heavy metals. World Health Organization, Geneva.

[65] Bernard A.M., Lauwerys R. (1986) Effects of cadmium exposure in humans, in: Foulkes E.C. (Ed.), Handbook of Experimental Pharmacology, Vol. 80. Springer Verlag, Berlin, pp. 135–77.

[66] Human Biomonitoring Commission of the German Federal Environmental Agency (2005) New and revised reference values for trace elements in blood and urine for children – arsenic, lead, cadmium, and mercury. Bundesgesundheitsbl, Gesundheitsforsch, Gesundheitsschutz, 48(11), 1308–1312.

[67] Tohyama C., Mitane Y., Kobayashi E. (1988) The relationships of urinary metallothionein with other indicators of renal dysfunction in people living in a cadmium-polluted area in Japan. Journal of Applied Toxicology, 8, 15–21.

[68] Boudreau J., Vincent R., Nadeau D., Trottier B., Fournier M., Krzystyniak K., Chevalier G. (1989) The response of the pulmonary surfactant-associated alkaline phosphatase following acute cadmium chloride inhalation. American Industrial Hygiene Association Journal, 50(7), 331–335.

[69] Piscator M. (1984) Long-term observations on tubular and glomerula function in cadmium-exposed persons. Environmental Health Perspectives, 54, 175–179.

[70] Kawada T., Tohyama C., Suzuki S. (1990) Significance of the excretion of urinary indicator proteins for a low level of occupational exposure to cadmium. International Archives of Occupational and Environmental Health, 62, 95–100.

[71] Lu J., Jin T., Nordberg G., Nordberg M. (2005) Metallothionein gene expression in peripheral lymphocytes and renal dysfunction in a population environmentally exposed to cadmium. Toxicology and Applied Pharmacology, 206, 150–156.

[72] International Agency for Research on Cancer (IARC) (1990). IARC monographs on the evaluation of carcinogenic risks to humans. Chromium, nickel and welding. Vol. 49. World Health Organization, Lyons, France, pp. 49–256.

[73] American Conference of Governmental Industrial Hygienists (ACGIH) (1999). Chromium. Documentation of the threshold limit values and biological exposure indices. American Conference of Governmental Industrial Hygienists. Cincinnati, OH, USA.

[74] Agency for Toxic Substances and Disease Registry (ATSDR) (1999). Toxicological Profile for Chromium. Agency for Toxic Substances and Disease Registry, Atlanta, GA, p. 421.

[75] Kuykendall J.R., Kerger B.D., Jarvi E.J. (1996) Measurement of DNA-protein cross-links in human leukocytes following acute ingestion of chromium in drinking water. Carcinogenesis, 17(9), 1971–1977.

[76] Agency for Toxic Substances and Disease Registry (ATSDR) (2004). Toxicological Profile for Cobalt. Department of Health and Human Services, Public Health Service, Atlanta, GA, USA.

[77] Lauwerys R., Lison D. (1994) Health risks associated with cobalt exposure – an overview. The Science of the Total Environment, 150, 1–6.

[78] Nemery B., Casier P., Roosels D., Lahaye D., Demedts M. (1992) Survey of cobalt exposure and respiratory health in diamond polishers. American Review of Respiratory Disease, 145, 610–616.

[79] Alimonti A., Bocca B., Mannella E., Petrucci F., Zennaro F., Cotichini R., D'Ippolito C., Agresti A., Caimi S., Forte G. (2005) Assessment of reference values for selected elements in a healthy urban population. Annali Istituto Superiore di Sanita, 41(2), 181–187.

[80] Al-Habsi K., Johnson E.H., Kadim I.T., Srikandakumar A., Annamalai K., Al-Busaidy R., Mahgoub O. (2007) Effects of low concentrations of dietary cobalt on liveweight gains, haematology, serum vitamin B(12) and biochemistry of Omani goats. Veterinary Journal, 173(1), 131–137.

[81] Shirakawa T., Kusaka Y., Fujimura N., Goto S., Kato M., Heki S., Morimoto K. (1989) Occupational asthma from cobalt sensitivity in workers exposed to hard metal dust. Chest, 95, 29–37.

[82] Potolicchio I., Festucci A., Hausler P., Sorrentino R. (1999) HLA-DP molecules bind cobalt: a possible explanation for the genetic association with hard metal disease. European Journal of Immunology, 29(7), 2140–2147.

[83] Agency for Toxic Substances and Disease Registry (ATSDR) (2005) Toxicological Profile for lead. Department of Health and Human Services, Public Health Service. Atlanta, GA, USA.

[84] Lim Y.C., Chia K.S., Ong H.Y. (2001) Renal dysfunction in workers exposed to inorganic lead. Annals Academy of Medicine Singapore, 30(2), 112–117.

[85] Anttila A., Heikkila P., Pukkala E. (1995) Excess lung cancer among workers exposed to lead. Scandinavian Journal of Work, Environment & Health, 21, 460–469.

[86] International Agency for Research on Cancer (IARC) (2004) Overall evaluations of carcinogenicity to humans: as evaluated in IARC Monographs Vols. 1–82 (at total of 900 agents, mixtures and exposures). Lyon, France. URL: http://www-cie.iarc.fr/monoeval/crthall.html.

[87] Scientific Committee on Occupational Exposure Limits for Lead and its Inorganic Compounds (2002) URL: http://ec.europa.eu/employment_social/ health_safety/docs/oel sum_83 lead_en.pdf.

[88] American Conference of Governmental Industrial Hygienists (ACGIH) (2001) Threshold limit values for chemical substances and physical agents and biological exposure indices. Cincinnati, OH, USA.

[89] Environmental Protection Agency (EPA) (1986) Air quality criteria for lead (EPA 600/8-83-028F Office of Research and Development, Environmental Criteria and Assessment Office), Research Triangle Park, NC, USA.

[90] Skerfving S. (1993) Inorganic lead, in: Beije B., Lundberg P., (Eds.), Criteria Documents from the Nordic Expert Group, Vol. 1. Nordic Council of Ministers, Copenhagen, DK, pp. 125–238.

[91] Wilhelm M., Pesch A., Rostek U., Begerow J., Schmitz N., Idel H., Ranft U. (2002) Concentrations of lead in blood, hair and saliva of German children living in three different areas of traffic density. The Science of the Total Environment, 297(1), 109–118.

[92] URL: http://www.cdc.gov/genomics/hugenet/file/print/factsheets/FS_ALAD. pdf.

[93] International Agency for Research on Cancer (IARC) (1987) IARC monographs on the evaluation of carcinogenic risk to humans. International Agency for Research on Cancer, World Health Organization, (Suppl. 7), 1–47.

[94] U.S. Environmental Protection Agency (1999) Integrated Risk Information System (IRIS) on Methyl Mercury. National Center for Environmental Assessment, Office of Research and Development, Washington, DC, USA.

[95] Deutsche Forschungsgemeinschaft (DFG) (1998) MAK- und BAT-Werte-Liste. Wiley-VCH, Weinheim, D.

[96] Nordberg G., Brune D., Gerhardsson L., Grandjean P., Vesterberg O., Wester P.O. (1992) The ICOH and IUPAC international programme for establishing reference values of metals. The Science of the Total Environment, 120(1–2), 17–21.

[97] Cherian M.G., Hursh J.B., Clarkson T.W., Allen J. (1978) Radioactive mercury distribution in biological fluids and excretion in human subjects after inhalation of mercury vapor. Archives of Environmental Health, 33, 109–114.

[98] Nielsen J.B., Andersen O. (1992) Transplacental passage and fetal deposition of mercury after low-level exposure to methylmercury – effect of seleno-L-methionine. Journal of Trace Elements and Electrolytes in Health and Disease, 6, 227–232.

[99] Cardenas A., Roels H., Bernard A.M., Barbon R., Buchet J.P., Lauwerys R.R. (1993) Markers of early renal changes induced by industrial pollutants. I Application to workers exposed to mercury vapour. Journal of Industrial Medicine, 50, 17–27.

[100] Stonard M.D., Chater B.V., Duffield D.P., Nevitt A.L., O'Sullivan J.J., Steel G.T. (1983) An evaluation of renal function in workers occupationally exposed to mercury vapor. International Archives of Occupational and Environmental Health, 52, 177–189.

[101] Langworth S., Elinder C.G., Sundquist K.G., Vesterberg O. (1992) Renal and immunological effects of occupational exposure to inorganic mercury. British Journal of Industrial Medicine, 49, 394–401.

[102] Strange R.C., Spiteri M.A., Ramachadran S., Fryer A.A. (2001) Glutathione-S-transferase family of enzymes. Mutation Research, 482, 21–26.

[103] Brambila E., Lientu J., Morgan D.L., Beliles R.P., Waalkes M.P. (2002) Effect of mercury vapor exposure on metallothionein and glutathione S-transferase gene expression in the kidney of nonpregnant, pregnant and neonatal rats. Journal of Toxicology and Environmental Health, 65, 1273–1288.

[104] Templeton D.M., Sunderman F.W., Herber F.M. (1994) Tentative reference values for nickel concentrations in human serum, plasma, blood, and urine: evaluation according to the TRACY protocol. Science and the Total Environment, 148, 243–251.

[105] Alimonti A., Petrucci F., Krachler M., Bocca B., Caroli S. (2000) Reference values for chromium, nickel and vanadium in urine of youngsters from the urban area of Rome. Journal of Environmental Monitoring, 2(4), 351–354.

[106] Angerer J., Lehnert G. (1990) Occupational chronic exposure to metals. II. Nickel exposure of stainless steal welders – biological monitoring. International Archives of Occupational and Environmental Health, 62, 7–10.

[107] Bernacki E.J., Parsons G.E., Roy B.R., Mikac-Devic M., Kennedy C.D., Sunderman F.W., (1978) Urine nickel concentrations in nickel-exposed workers. Annals of Clinical and Laboratory Science, 8 (3), 184–189.

[108] Elias Z., Mur J.M., Pierre F., Gilgenkrantz S., Schneider O., Baruthio F., Danière M.C., Fontana J.M. (1989) Chromosome aberrations in peripheral blood lymphocytes of welders and characterization of their exposure by biological samples analysis. Journal of Occupational Medicine, 31(5), 477–483.

[109] Ghezzi A., Baldasseroni G., Sesana C., Boni G., Alessio L. (1989) Behaviour of urinary nickel in low-level occupational exposure. Giornale Italiano di Medicina del Lavoro ed Ergonomia, 80, 244–250.

[110] Sunderman F.W. Jr. (1993) Biological monitoring of nickel in humans. Scandinavian Journal of Work and Environmental Health, 19(Suppl. 1), 34–38.

[111] Frenkel K., Karkoszkan J., Cohen B., Baranski B., Jajubowski M., Cosma G., Taioli E., Toniolo P. (1994) Occupational exposure to Cd, Ni, and Cr modulate titers of antioxidized DNA base antibodies. Environmental Health Perspectives, 102, (Suppl. 3), 221–225.

[112] Reith A.K., Reichborn-Kenneruud S., Aubele M., Ytting U.J., Gais P., Burger G. (1994) Biological monitoring of chemical exposure in nickel workers by imaging cytometry (ICM) of nasal smears. Analytical Cellular Pathology, 6, 9–21.

[113] Mozzanica N., Rizzolo L., Veneroni G. (1990) HLA-A, B, C and DR antigens in nickel contact sensitivity. British Journal of Dermatology, 122, 309–314.

[114] Agency for Toxic Substances and Disease Registry (ATSDR) (2007) Toxicological Profile for Thallium. Department of Health and Human Services, Public Health Service, Atlanta, GA, USA.

[115] Schaller K.H., Manke G., Raithel H.J., Buhlmeyer G., Schmidt M., Valentin H. (1980) Investigations of thallium-exposed workers in cement factories. International Archives of Occupational and Environmental Health, 47(3), 223–231.

[116] Brockhaus A., Dolgner R., Ewers U., Dramer U., Soddemann H., Wiegland H. (1981) Intake and health effects of thallium among a population living in the vicinity of a cement plant emitting thallium containing dust. International Archives of Occupational and Environmental Health, 48, 375–389.

[117] Apostoli P., Maranelli G., Minoia C., et al. (1988) Urinary thallium: critical problems, reference values and preliminary results of an investigation in workers with suspected industrial exposure. The Science of the Total Environment, 71, 513–518.

[118] Gastel B. (Ed.) (1978) Thallium poisoning. Johns Hopkins Medical Journal, 142, 27–31.

[119] Environmental Protection Agency (EPA) (1979) URL: http://www.epa.gov/opptintr/pcb/.

[120] Wang S.L., Chang Y.C., Chao H.R., Li C.M., Li L.A., Lin L.Y., Papke O. (2006) Body burdens of polychlorinated dibenzo-*p*-dioxins, dibenzofurans, and biphenyls and their relations to estrogen metabolism in pregnant women. Environmental Health Perspectives, 114(5), 740–745.

[121] Iida T., Todaka T., Hirakawa H., Tobiishi K., Matsueda T., Hori T., Nakagawa R., Furue M. (2003) Follow-up survey of dioxins in the blood of Yusho patients (in 2001). Fukuoka Igaku Zasshi – Fukuoka Acta Medica, 94(5), 126–135.

[122] Kashimoto T., Miyata H., Fukushima S., Kunita N., Ohi G., Tung T.C. (1985) PCBSs, PCQs and PCDFs in blood of yusho and yu-cheng patients. Environmental Health Perspectives, 59, 73–78.

[123] Chen P.H., Wong C.K., Rappe C., Nygren M. (1985) Polychlorinated biphenyls, dibenzofurans and quaterphenyls in toxic rice-bran oil and in the blood and tissues of patients with PCBS poisoning (Yu-Cheng) in Taiwan. Environmental Health Perspectives, 59, 59–65.

[124] Masuda Y., Kuroki H., Haraguchi K., Nagayama J. (1985) PCBS and PCDF congeners in the blood and tissues of yusho and yu-cheng patients. Environmental Health Perspectives, 59, 53–58.

[125] Agency for Toxic Substances and Disease Registry (ATSDR) (2000) Toxicological Profile for Polychlorinated Biphenyls. U.S. Dept. of Health and Human Services, Public Health Service, Washington, DC, p. 765.

[126] Carpenter D.O. (2006) Polychlorinated biphenyls (PCBs): routes of exposure and effects on human health. Reviews on Environmental Health, 21(1), 1–23.

[127] McKinney J.D., Waller C.L. (1994) Polychlorinated biphenyls as hormonally active structural analogues. Environmental Health Perspectives, 102(3), 290–297.

[128] Kilic N., Sandal S., Colakolu N., Kutlu S., Seyran A., Yilmaz B. (2005) Endocrine disruptive effects of polychlorinated biphenyls on the thyroid gland in female rats. Tohoku Journal of Experimental Medicine, 206(4), 327–332.

[129] Maroni M., Colombi A., Arbosti G., Cantoni S., Foa V. (1981) Occupational exposure to polychlorinated biphenyls in electrical workers. II. Health effects. British Journal of Industrial Medicine, 38(1), 55–60.

[130] Larsen J.C. (2006) Risk assessments of polychlorinated dibenzo-p-dioxins, polychlorinated dibenzofurans, and dioxin-like polychlorinated biphenyls in food. Molecular Nutrition & Food Research, 50(10), 885–896.

[131] Jacobson J.L., Jacobson S.W., Humphrey H.E. (1990) Effects of exposure to PCBs and related compounds on growth and activity in children. Neurotoxicology and Teratology, 12(4), 319–326.

[132] Nicholson W.J., Landrigan P.J. (1994) Human health effects of polychlorinated biphenyls, in: Schecter A. (Ed.), Dioxins and Health. Plenum, New York, pp. 487–524.

[133] Negri E., Borsetti C., Fattore E., La Vecchia C. (2003) Environmental exposure to polychlorinated biphenyls (PCBs) and breast cancer: a systematic review of the epidemiological evidence. European Journal of Cancer Prevention, 12(6), 509–516.

[134] Chen G., Bunce N.J. (2004) Interaction between halogenated aromatic compounds in the Ah receptor signal transduction pathway. Environmental Toxicology, 19(5), 480–489.

[135] Kerkvliet N.I. (1995) Immunological effects of chlorinated dibenzo-p-dioxins. Environmental Health Perspectives, 103(Suppl. 9), 47–53.

[136] Vondracek J., Machala M., Bryja V., Chramostova K., Krcmar P., Dietrich C., Hampl A., Kozubik A. (2005) Aryl hydrocarbon receptor-activating polychlorinated biphenyls and their hydroxylated metabolites induce cell proliferation in contact-inhibited rat liver epithelial cells. Toxicological Sciences, 83(1), 53–63.

[137] Furst P. (2006) Dioxins, polychlorinated biphenyls and other organohalogen compounds in human milk. Levels, correlations, trends and exposure through breastfeeding. Molecular Nutrition & Food Research, 50(10), 922–933.

[138] OSHA (1998) Occupations Safety and Health Standards, Air Contaminants. 29 CFR 1910.1000.

[139] Giesy J.P., Kannan K. (1998) Dioxin-like and non-dioxin-like toxic effects of polychlorinated biphenyls (PCBs): implications for risk assessment. Critical Reviews in Toxicology, 28(6), 511–569.

[140] Wolff M.S., Britton J.A., Teitelbaum S.L., Eng S., Deych E., Ireland K., Liu Z., Neugut A.I., Santella R.M., Gammon M.D. (2005) Improving organochlorine biomarker models for cancer research. Cancer Epidemiology, Biomarkers & Prevention, 14(9), 2224–2236.

[141] Fait A., Grossman E., Self S., Jeffries J., Pellizzar E.D., Emmett E.A. (1989) Polychlorinated biphenyl congeners in adipose tissue lipid and serum of past and present transformer repair workers and a comparison group1. Toxicological Sciences, 12(1), 42–55.

[142] DeKoning E.P., Karmaus W. (2000) PCBS exposure in utero and via breast milk. A review. Journal of Exposure Analysis and Environmental Epidemiology, 10(3), 285–293.

[143] Longnecker M.P., Rogan W.J., Lucier G. (1997) The human health effects of DDT (dichlorodiphenyltrichloroethane) and PCBSs (polychlorinated biphenyls) and an overview of organochlorines in public health. Annual Review of Public Health, 18, 211–244.

[144] Guo Y.L., Yu M.L., Hsu C.C., Rogan W.J. (1999) Chloroacne, goiter, arthritis, and anemia after polychlorinated biphenyl poisoning: 14-year follow-up of the Taiwan Yucheng cohort. Environmental Health Perspectives, 107(9), 715–719.

[145] Fitzgerald E.F., Hwang S.A., Lambert G., Gomez M., Tarbell A. (2005) PCBS exposure and in vivo CYP1A2 activity among native Americans. Environmental Health Perspectives, 113(3), 1–6.

[146] Nims R.W., Fox S.D., Issaq H.J., Lubet, R.A. (1992) Accumulation and persistence of individual polychlorinated biphenyl congeners in liver, blood, and adipose tissue of rats following dietary exposure to Aroclor® 1254. Archives of Environmental Contamination and Toxicology, 27(4), 413–520.

[147] Alvares A.P., Fischbein A., Anderson K.E. (1977) Alterations in drug metabolism in workers exposed to polychlorinated biphenyls. Clinical Pharmacology & Therapeutics, 22, 140–146.

[148] Miranda S.R., Meyer S.A. (2007) Cytotoxicity of chloroacetanilide herbicide alachlor in HepG2 cells independent of CYP3A4 and CYP3A7. Food and Chemical Toxicology, 45(5), 871–877.

[149] Calabrese E.J., Sorenson A.J. (1977) The health effects of PCBs with particular emphasis on human high risk groups. Reviews on Environmental Health, 2, 285–304.

[150] Rowse D.H., Emmett E.A. (2004) Solvents and the skin. Clinics in Occupational & Environmental Medicine, 4(4), 657–730.

[151] Szeszenia-Dabrowska N., Wilczyska U., Kaczmarek T., Szymczak W. (1991) Cancer mortality among male workers in the Polish rubber industry. Polish Journal of Occupational Medicine & Environmental Health, 4(2), 149–57.

[152] Iwata T., Mori H., Dakeishi M., Onozaki I., Murata K. (2005) Effects of mixed organic solvents on neuromotor functions among workers in Buddhist altar manufacturing factories. Journal of Occupational Health, 47(2), 143–148.

[153] Bode J.C., Kuhn C. (1992) Liver damage from organic solvents. Deutsche Medizinische Wochenschrift, 117(28–29), 1127–1129.

[154] Stefankiewicz J., Kurzawa R., Drozdzik M. (2006) Environmental factors disturbing fertility of men. Ginekologia Polska, 77(2), 163–169.

[155] Lamb J.C., Hentz K.L. (2006) Toxicological review of male reproductive effects and trichloroethylene exposure: assessing the relevance to human male reproductive health. Reproductive Toxicology, 22(4), 557–563.

[156] Thulstrup A.M., Bonde J.P. (2006) Maternal occupational exposure and risk of specific birth defects. Occupational Medicine, 56(8), 532–543.

[157] Brautbar N., Wu M.P., Gabel E., Regev L. (2006) Occupational kidney cancer: exposure to industrial solvents. Annals of the New York Academy of Sciences, 1076, 753–764.

[158] Hayes R.B. (1992) Biomarkers in occupational cancer epidemiology: considerations in study design. Environmental Health Perspectives, 98, 149–154.

[159] Agency for Toxic Substances and Disease Registry (ATSDR) (1997) Toxicological profile for benzene, U.S. Public Health Service, U.S. Department of Health and Human Services, Atlanta, GA.

[160] Bloemen L.J., Youk A., Bradley T.D., Bodner K.M., Marsh G. (2004) Lymphohaematopoietic cancer risk among chemical workers exposed to benzene. Occupational and Environmental Medicine, 61, 270–274.

[161] Snyder R., Hedii C.C. (1996) An overview of benzene metabolism. Environmental Health Perspectives, 104(Suppl. 6), 1165–1171.

[162] Bodell W.J., Levay G., Pongracz K. (1993) Investigation of benzene-DNA adducts and their detection in human bone marrow. Environmental Health Perspectives, 99, 241–244.

[163] Rothman N., Bechtold W.E., Yin S.-N., Dosemeci M., Li G.-L., Wang Y.-Z., Griffith W.C., Smith M.T., Hayes R.B. (1998) Urinary excretion of phenol, catechol, hydroquinone, and muconic acid by workers occupationally exposed to benzene. Occupational and Environmental Medicine, 55, 705–711.

[164] Roush G.J., Ott M.G. (1985) A study of benzene exposure versus urinary phenol levels. American Industrial Hygiene Association Journal, 7, 385–393.

[165] Stommel P., Muller G., Stucker W., Verkoyen C., Schobel S., Norpoth K. (1989) Determination of S-phenylmercapturic acid in the urine – an improvement in the biological monitoring of benzene exposure. Carcinogenesis, 10(2), 279–282.

[166] Scheepers P.T.J., Heussen G.A.H. (2005) New and improved biomarkers ready to be used in health-risk oriented exposure and susceptibility assessments: report of the 6th International Symposium on Biological Monitoring in Occupational and Environmental Health, Biomarkers, 10(1), 80–94.

[167] EPA (2002) Toxicological review of benzene. U.S. Environmental Protection Agency, Washington, DC, p. 180.

[168] Agency for Toxic Substances and Disease Registry (ATSDR) (2000) Toxicological profile for toluene, U.S. Public Health Service, U.S. Department of Health and Human Services, Atlanta, GA.

[169] Andersen I., Lundqvist G.R., Molhave L. (1983) Human response to controlled levels of toluene in six-hour exposures. Scandinavian Journal of Work, Environment and Health, 9, 405–418.

[170] Shibata K., Yoshita Y., Matsumoto H. (1994) Extensive chemical burns from toluene. American Journal of Emerging Medicine, 12(3), 353–355.

[171] Kamijo Y., Soma K., Hasegawa I. (1998) Fatal bilateral adrenal hemorrhage following acute toluene poisoning: a case report. The Journal of Toxicology – Clinical Toxicology, 36(4), 365–368.

[172] Burgaz S., Erdem O., Cakmak G., Erdem N., Karakaya A., Karakaya A.E. (2002) Cytogenetic analysis of buccal cells from shoe-workers and pathology and anatomy laboratory workers exposed to n-hexane, toluene, methyl ethyl ketone and formaldehyde. Biomarkers 7, 151–161.

[173] Tassaneeyakul W., Birkett D.J., Edwards J.W. (1996) Human cytochrome P450 isoform specificity in the regioselective metabolism of toluene and o-, and p-xylene. Journal of Pharmacology and Experimental Therapeutics, 276(1), 101–108.

[174] Nakajima T., Wang R.-S., Elovaara E. (1997) Toluene metabolism by cDNA-expressed human hepatic cytochrome P-450. Biochemical Pharmacology, 53(3), 271–277.

[175] Kawai T., Yasugi T., Mizunuma K. (1992) Comparative evaluation of urine analysis and blood analysis as means of detecting exposure to organic solvents at low concentrations. International Archives of Occupational and Environmental Health, 64, 223–224.

[176] Maestri L., Ghittori S., Imbriani M. (1997) Determination of specific mercapturic acids as an index of exposure to environmental benzene, toluene, and styrene. Industrial Health, 35, 489–501.

[177] Lof A., Hjelm E.W., Colmsjo A. (1993) Toxicokinetics of toluene and urinary excretion of hippuric acid after human exposure to 2H8-toluene. British Journal of Industrial Medicine, 50, 55–59.

[178] Inoue O., Seiji K., Watanabe T. (1986) Possible ethnic difference in toluene metabolism: a comparative study among Chinese, Turkish and Japanese solvent workers. Toxicology Letters, 34, 167–174.

[179] Kawamoto T., Koga M., Oyama T. (1996) Habitual and genetic factors that affect urinary background levels of biomarkers for organic solvent exposure. Archives of Environmental Contamination and Toxicology, 30(1), 114–120.

[180] Angerer J., Schildbach M., Kramer A. (1998) S-*p*-toluylmercapturic acid in the urine of workers exposed to toluene: a new biomarker for toluene exposure. Archives of Toxicology, 72(2), 119–123.

[181] ACGIH (1999) TLVs and BEIs: Threshold Limit Values for Chemical Substances and Physical Agents: Biological Exposure Indices. American Conference of Governmental Industrial Hygienists, Cincinnati, OH.

[182] Wiwanitkit V., Suwansaksri J., Soogarun S. (2006) White blood cell sister chromatid exchange among a sample of Thai subjects exposed to toluene, an observation. The International Journal of Experimental Pathology, 87, 501–503.

[183] Mendoza-Cantú A., Castorena-Torres F., Bermúdez de León M., Cisneros B., López-Carrillo E., Rojas-García A.E., Aguilar-Salinas A., Manno M., Albores A. (2006) Occupational toluene exposure induces cytochrome P450 2E1 mRNA expression in peripheral lymphocytes. Environmental Health Perspectives, 114(4), 494–499.

[184] Holian A. (1996) Air toxics: biomarkers in environmental applications – overview and summary of recommendations. Environmental Health Perspectives, 104(Suppl. 5), 851–855.

[185] Eddleston M., Karalliede L., Buckley N., Fernando R., Hutchinson G., Isbiter G., Konradsen F., Murray D., Piola J.C., Senanayake N., Sheriff R., Singh S., Siwach S.B., Smit L. (2002) Pesticide poisoning in the developing world, a minimum pesticides list. Lancet, 360, 1163–1167.

[186] Al-Saleh I.A. (1994) Pesticides: a review article. The Journal of Environmental Pathology, Toxicology and Oncology, 13(3), 151–61.

[187] WHO/UNEP (1990) Public Health Impact of Pesticides Used in Agriculture. World Health Organization, Geneva.

[188] IARC (1991) Occupational exposures in insecticide application, and some pesticides, in: IARC Monographs on the Evaluation of the Carcinogenic Risk of Chemicals to Humans. Vol. 53. International Agency for Research on Cancer, Lyon.

[189] Restrepo M., Munoz N., Day N., Parra J.E., Hernandez C., Blettner M., Giraldo A. (1990) Birth defects among children born to a population occupationally exposed to pesticides in Colombia. Scandinavian Journal of Work, Environment & Health, 16, 239–246.

[190] Blair A., Grauman D.J., Lubin J.H., Fraumeni J.F. (1983) Lung cancer and other causes of death among licensed pesticide applicators. Journal of the National Cancer Institute, 71, 31–37.

[191] MacMahon B., Monson R.R., Wang H.H., Tongzhang Z. (1988) A second follow-up of mortality in a cohort of pesticide applicators. Journal of Occupational Medicine, 30, 429–432.

[192] Barthel E. (1981) Increased risk of lung cancer in pesticide-exposed male agricultural workers. Journal of Toxicology and Environmental Health, 8, 745–748.

[193] Corrao G., Calleri M., Carle F., Russo R., Bosia S., Piccioni P. (1989) Cancer risk in a cohort of licensed pesticide users. Scandinavian Journal of Work, Environment & Health, 15, 203–209.

[194] Hoar Zahm S., Blair A., Holmes F.F., Boysen C.D., Robel R.J. (1988) A case-referent study of soft-tissue sarcoma and Hodgkin's disease. Farming and insecticide use. Scandinavian Journal of Work, Environment & Health, 14(4), 224–230.

[195] Woods J.S., Polissar L. (1989) Non-Hodgkin's lymphoma among phenoxy herbicide-exposed farm workers in western Washington State. Chemosphere, 18, 401–406.

[196] Coye M.J., Lowe J.A., Maddy K.T. (1986) Biological monitoring of agricultural workers exposed to pesticides: I. Cholinesterase activity determinations. Journal of Occupational Medicine, 28, 619–627.

[197] NCHS (2003) National Health and Nutrition Examination Survey. Hyattsville, MD: National Center for Health Statistics. URL: http://www.cdc.gov/nchs/nhanes.htm.

[198] Bertram H.P., Kemper F.H., Muller C. (1986) Hexachlorobenzene content in human whole blood and adipose tissue: experiences in environmental specimen banking. IARC Scientific Publications, 77, 173–183.

[199] Agency for Toxic Substances and Disease Registry (ATSDR) (2000) Toxicological Profile for Hexachlorobenzene. Update. (Draft for Public Comment). Agency for Toxic Substances and Disease Registry, Atlanta, GA, p. 349.

[200] Agency for Toxic Substances and Disease Registry (ATSDR) (2005) Toxicological Profile for Alpha-, Beta-, Gamma-, and Delta-Hexachlorocyclohexane. (Update). U.S. Department of Public Health and Human Services, Public Health Service, Atlanta, GA.

[201] Health Council of the Netherlands: Dutch Expert Committee on Occupational Standards (2001) Lindane (γ-hexachlorocyclohexane); Health-based recommended occupational exposure limit. Health Council of the Netherlands, publication no. 2001/07OSH.

[202] Hryhorczuk D.O., Moomey M., Burton A., Runkle K., Chen E., Saxer T., Slightom J., Dimos J., McCann K., Barr D. (2002) Urinary p-nitrophenol as a biomarker of household exposure to methyl parathion. Environmental Health and Perspectives, 110(Suppl. 6), 1041–1046.

[203] Dengan N., Moldeus P., Kasilo O.M.J., Nhachi C.F.B. (1995) Use of urinary p-nitrophenol as an index of exposure to parathion. Bulletin of Environmental Contamination and Toxicology, 55(2), 296–302.

[204] Abu-Qare A., Abou-Donia M.B. (2000) Urinary excretion of metabolites following a single dermal dose of [^{14}C] methyl parathion in pregnant rats. Toxicology, 150, 119–127.

[205] Abu-Qare A.W., Abdel-Rahman A.A., Ahmad H., Kishk A.M., Abou-Donia M.B. (2001) Absorption, distribution, metabolism and excretion of daily oral doses of [^{14}C] methyl parathion in hens. Toxicology Letters, 125, 1–10.

[206] Jordan L.W., Neal R.A. (1976) Examination of the in vivo metabolism of maneb and zineb to ethylene-thiourea (ETU) in mice. Bulletin of Environmental Contamination and Toxicology, 22, 271–277.

[207] IARC (1987) Overall evaluations of carcinogenicity: an updating of IARC Monographs. Vols. 1–42. IARC Monogr Eval Carcinog Risks Hum. International Agency For Research on Cancer, Lyon, Suppl. 7, p. 207.

[208] IARC (2001) Some thyrotropic agents. IARC Monogr Eval Carcinog Risks Hum. International Agency For Research on Cancer, Lyon, 79, p. 659.

[209] Aprea C., Betta A., Catenacci G., Lotti A., Minoia C., Passini W., Pavan I., Robustelli della Cuna F.S., Roggi C., Ruggeri R., Soave C., Sciarra G., Vannini P., Vitalone V. (1996) Reference values of urinary ethylenethiourea in four regions of Italy (multicentric study). The Science of the Total Environment, 192, 83–93.

[210] Walker C.H. (1992) Biochemical responses as indicators of toxic effects of chemicals in ecosystems. Toxicology Letters, 64–65, 527–533.

[211] Ames R.G., Brown S.K., Mengle D.C., Kahn E., Stratton J.W., Jackson R.J. (1989) Cholinesterase activity depression among California agricultural pesticide applicators. American Journal of Industrial Medicine, 15, 143–150.

[212] McConnell R., Magnotti R. (1994) Screening for insecticide overexposure under field conditions: a reevaluation of the tintometric cholinesterase kit. American Journal of Public Health, 84, 479–81.

[213] Sultatos L.G. (1994) Mammalian toxicology of organophosphorus pesticides. Journal of Toxicology and Environmental Health, 43, 271–289.

[214] Sulbatos L.G. (1994) Mammalian toxicology of organophosphorous pesticides. Journal of Toxicology and Environmental Health, 43, 271–289.

[215] Guengerichi F.P., Kim D.H., Iwasaki M. (1991) Role of human cytochrome P450 IIE1 in the oxidation of many low molecular weight cancer suspects. Chemical Research in Toxicology, 4, 168–179.

[216] Uematsu F., Kikuchi H., Motomya M., Abe T., Sagami I., Ohmachi T., Wakui A., Kanamaru R., Watanabe M. (1991) Association between restriction fragment length polymorphism of the human cytochrome P450 2E1 gene and susceptibility to lung cancer. Japanese Journal of Cancer Research, 82, 254–256.

[217] Coles B., Ketterer B. (1990) The role of glutathione and glutathione transferases in chemical carcinogenesis. Critical Reviews in Biochemistry and Molecular Biology, 25, 47–70.

[218] Costa L.G., Li W.F., Richter R.J., Shih D.M., Lusis A., Furlog C.E. (1999) The role of paraoxonase (PON1) in the detoxication of organophosphates and its human polymorphism. Chemico-Biological Interactions, 119, 429–438.

[219] Furlog C.E., Li W.F., Brophy V.H., Jarvik G.P., Richter R.J., Shih D.M., Lusis A.J., Costa L.G. (2000) The PON1 gene and detoxication. Neurotoxicology, 21, 581–587.

[220] Clayson D.B., Garner R.C. (1976) Carcinogenic aromatic amines and related compounds, in: Searle C.E. (Ed.), Chemical Carcinogens. ACS Monograph 173, American Chemical Society, Washington, DC, pp. 36–461.

[221] Case R.A.M., Hosker M.E., McDonald D.B., Pearson J.T. (1954) Tumours of the urinary bladder in workmen engaged in the manufacture and use of certain dyestuff intermediate in the British chemical industry. Part I. The role of aniline, benzidine, alpha-naphthylamine, and beta-naphthylamine. British Journal of Industrial Medicine, 11, 75–104.

[222] International Agency for Research on Cancer (IARC) (1978) IARC Monograph on the Evaluation of the Carcinogenic Risk of Chemicals to Humans. Vol. 17, Some N-Nitroso Compounds. International Agency for Research on Cancer, Lyons, p. 365.

[223] Palmiotto G., Pieraccini G., Moneti G., Dolora P. (2001) Determination of the levels of aromatic amines in indoor and outdoor air in Italy. Chemosphere, 43(3), 355–361.

[224] Pavan I., Buglione E., Colombi A., Cottica D., Della Vedova L., Larsen B., Micoli G., Mosconi G., Passini V. (1994) Aspetti metodologici e concettuali nella definizione di valori di riferimento di ammine aromatiche in fluidi biologici. ed. Morgan.

[225] American Conference of Governmental Industrial Hygienists (ACGIH). (1999) TLVs and BEIs. Threshold Limit Values for Chemical Substances and Physical Agents. Biological Exposure Indices, Cincinnati, OH.

[226] Bryant M.S., Vineis P., Skipper P.L., Tannenbaum S.R. (1988) Hemoglobin Adducts of Aromatic Amines: Associations with Smoking Status and Type of Tobacco. Proceedings of the National Academy of Sciences of the United States of America, 85(24), 9788–9791.

[227] Snyderwine E.G. (1994) Some perspectives on the nutritional aspects of breast cancer research: food-derived heterocyclic amines as etiologic agents in human mammary cancer. Cancer, 74, 1070–1077.

[228] Lindsay M.M., Schenkb M., Heinc D.W., Davisd S., Zahma S.H., Cozene W., Cerhanf J.R., Hartgea P., Welchg R., Chanockg S.J., Rothmana N., Wan S.S. (2006) Genetic variation in N-acetyltransferase 1 (NAT1) and 2 (NAT2) and risk of non-Hodgkin lymphoma. Pharmacogenetics and Genomics, 16(8), 537–545.

[229] OSH (1982) The Health Consequences of Smoking, Cancer, A Report of the Surgeon General. DHHS (PHS).

[230] HSDB (2001) Hazardous Substances Data Base. National Library of Medicine. URL: http://toxnet.nlm.nih.gov/cgi-bin/sis/htmlgen?HSDB.

[231] NIOSH (1984) National Occupational Exposure Survey (1981–83). Cincinnati, OH: U. S. Department of Health and Human Services. URL: http://www.cdc.gov/noes/noes3/empl0003.htm.

[232] Rubino F., Scansetti G., Piolatto G., Pira E. (1982) The carcinogenic effect of aromatic amines: an epidemiological study on the role of o-toluidine and 4,4'-methylene bis (2-methylaniline) in inducing bladder cancer in man. Environmental Research, 27, 241–254.

[233] Ward E., Carpenter A., Markowitz S., Roberts D.W. (1991) Excess numbers of bladder cancers in workers exposed to *ortho*-toluidine and aniline. Journal of the National Cancer Institute, 83, 501–506.

[234] Markowitz S.B., Levin K. (2004) Continued epidemic of bladder cancer in workers exposed to *ortho*-toluidine in a chemical factory. Journal of Occupational and Environmental Medicine, 46, 154–160.

[235] IARC (2000) Some industrial chemicals. IARC Monographs on the Evaluation of Carcinogenic Risks to Humans, Vol. 77. International Agency for Research on Cancer, Lyon, France, p. 564.

[236] Deutsche Forschungsgemeinschaft (2006) Deutsche Forschungsgemeinschaft, MAK- und BAT-Werte-Liste. Mitteilung 42 der Senats-kommission zur Prüfung gesundheitsschädlicher Arbeitsstoffe, Wiley/VCH Verlag GmbH & Co.KGaA, Weinheim, Germany.

[237] Skipper P.L., Tannenbaum S.R. (1990) Protein adducts in the molecular dosimetry of chemical carcinogens. Carcinogenesis, 11, 507–518.

[238] Richter E., Branner B. (2002) Biomonitoring of exposure to aromatic amines: haemoglobin adducts in humans. Journal of Chromatography B, 778(1), 49–62.

[239] Ward E.M., Sabbioni G., DeBord D.G., Teass A.W., Brown K.K., Talaska G.G., Roberts D.R., Ruder A.M., Streicher R.P. (1996) Monitoring of aromatic amine exposures in workers at a chemical plant with a known bladder cancer excess. Journal of National Cancer Institute, 88, 1046–1052.

[240] Jodynis-Liebert J., Matuszewska A. (1999) Effect of toluidines and dinitro-toluenes on caffeine metabolic rate in rat. Toxicology Letters, 104, 159–165.

[241] Benowitz N.L., Peng M.P. (2003) Effects of cigarette smoking and carbon monoxide on chlorzoxazone and caffeine metabolism. Clinical Pharmacology & Therapeutics, 74, 468–474.

7 Multivariate approaches to biomonitoring studies

M E Conti & M Mecozzi

1 Introduction

The statistical interpretation of experimental data in environmental analysis is a fundamental step which can be performed by two different techniques based on univariate and multivariate approaches, respectively. The use of the univariate approach is generally more common due to several reasons. The mathematical tools necessary for the univariate approach are generally simple and unless large data sets many tests can be also performed by hand calculators; in addition, many common functions of univariate methods are implemented on spreadsheets for PC. This approach was fully discussed elsewhere [1].

The mathematical tools of multivariate analysis are more complex because they are based on matrix algebra so that they require specific software packages. However, the multivariate approach has some peculiar advantages with respect to the univariate approach. In fact, it allows to examine the data taking into account the global contribution of all the variables present in the data set. With respect to the univariate approach, this allows to highlight better the statistical weight of each variable in the data distribution, showing similarities and differences among samples with many details [2].

Multivariate methods include several techniques which are used with different aims. Some techniques are used to perform a preliminary exploratory data analysis when no one specific information is known about the general characteristics of data and their distribution; for this reason they are defined "unsupervised pattern recognition" methods. Other techniques are applied to perform the differentiation of data samples when we wish to identify peculiar characteristics (i.e. variables) which can determine a separation of data in two or more typologies (i.e. classes). As they are used after a preliminary exploration of data performed by one of the techniques belonging to the "unsupervised pattern recognition" methods, they are defined "supervised pattern recognition" methods.

Recently, due to the diffusion of high efficiency personal computers and to availability of several statistical packages, the complexity of mathematical tools involved in multivariate analysis can be overcome. In addition, recent versions of common spreadsheets such as Excel for Windows include many routines of matrix algebra so that some multivariate tests can be performed even without the availability of specific statistical software.

In this chapter we describe the use of some multivariate methods belonging to unsupervised and supervised pattern recognition families in a practical case of biomonitoring study, related to the accumulation properties of heavy metals in two molluscs *Monodonta turbinata* and *Patella caerulea*.

2   Unsupervised pattern recognition: theory and application of principal component analysis (PCA)

PCA is a mathematical technique which allows to describe a data set with a lower number of variables than the original variables and for this reason it has been defined data reduction form [3, 4]. For a given data matrix $\mathbf{M}$ with dimension $r \times v$ where "r" is the number of samples (i.e. rows) in the set and "v" is the number of variables (i.e. columns) considered for each set, PCA decomposes $\mathbf{M}$ into the product of three new matrices

$$\mathbf{M} = \mathbf{SAL}' \tag{1}$$

$\mathbf{S}$ is the so-called score matrix having dimension $r \times f$ where r is the original number of samples and f is the number of the new variables called factors which include all the information previously described by the original v-variables of the $\mathbf{M}$ matrix but f is always significantly lower than v.

$\mathbf{A}$ is the so-called eigenvalues matrix having the same dimension of the $\mathbf{M}$ matrix, whose diagonal describes the portion of variance explained by each significant factor.

$\mathbf{L}'$ is the transpose of the so-called loading matrix $\mathbf{L}$ having dimension $v \times f$ and it describes the statistical significance (i.e. ability) of each original v-variable to describe the information of the $\mathbf{M}$ matrix.

Several excellent text books [3, 5] and papers [2, 6] describe the mathematical steps of $\mathbf{M}$ decomposition while an algorithm in MATLAB language for decomposing $\mathbf{M}$ has been reported by Geladi [7]. However, PCA routines are always included in several statistical software packages such as SPSS (http://www.spss.com/spss/), SAS (http://www.sas.com/), Statistica, (http://www.statsoft.com), PARVUS (http://www.parvus.unige.it/index.php) and in the chemometrical toolbox of some programming languages like MATLAB (http://www.mathworks. com).

Let us have a look at a practical example related to a biomonitoring study concerning the data reported in table 1; they represent the $\mathbf{M}$ matrix. Data are referred to trace metals determined in mollusc samples collected at Linosa Island, Sicily, Italy; for the experimental procedure see Conti *et al.* [8], Conti *et al.* [9].

Before PCA application the data are standardised according to a procedure consisting in the determination of the mean value "m" and standard deviation "s"

Table 1: Data set from Linosa Island, Sicilian channel, South Italy.

Sampling station	Ind.	Cd	Cr	Cu	Pb	Zn
Patella caerulea[a]						
1	8	3.25 ± 0.86	0.27 ± 0.88	2.78 ± 0.88	0.40 ± 0.11	41.78 ± 10.13
2	8	3.68 ± 1.01	0.24 ± 0.10	6.22 ± 1.61	0.47 ± 0.15	38.40 ± 6.26
3	8	3.26 ± 0.97	0.75 ± 0.22	4.04 ± 1.35	1.28 ± 0.31	41.47 ± 6.87
4	8	2.69 ± 0.63	0.72 ± 0.21	9.71 ± 2.25	2.41 ± 0.65	42.84 ± 4.44
5	8	4.97 ± 1.31	0.13 ± 0.04	6.60 ± 1.14	0.51 ± 0.09	51.47 ± 9.62
Monodonta turbinata[b]						
1	8	0.77 ± 0.23	0.45 ± 0.10	10.13 ± 2.81	1.66 ± 0.63	50.21 ± 2.27
2	8	1.17 ± 0.28	0.38 ± 0.08	15.31 ± 3.82	0.71 ± 0.22	50.47 ± 6.23
3	8	1.11 ± 0.20	1.30 ± 0.38	34.89 ± 9.78	1.07 ± 0.25	60.90 ± 8.63
4	8	1.11 ± 0.32	0.94 ± 0.28	28.26 ± 4.90	1.78 ± 0.49	59.29 ± 4.89
5	8	1.28 ± 0.35	0.15 ± 0.04	12.30 ± 2.37	0.62 ± 0.13	56.26 ± 8.10
Seawater ($n = 12$) Dissolved metals ($\mu g\ L^{-1}$)		0.16 ± 0.04	0.18 ± 0.02	0.96 ± 0.08	0.68 ± 0.15	10.66 ± 1.11

Concentrations of metals ($\mu g\ g^{-1}$ d.w.) in the soft tissues of *P. caerulea* and *M. turbinata* (means ± SD) and mean dissolved metal concentrations in seawater (means ± SD).

of each "v" variable according to

$$v_i' = \frac{(v_i - m_i)}{s_i} \qquad (2)$$

where v_i' is the new value of the v_i original value and m_i and s_i are the mean value and the standard deviation value of the ith variable.

A practical example of this procedure is reported in table 2. This pretreatment is always necessary when the variables cover different orders of magnitude [10]. A routine for data standardisation is generally present in any multivariate package. After standardisation, PCA is applied and the standardised **M** matrix is decomposed into the score matrix **S** (table 3) and in the loading matrix **L** (table 4).

The explained variance by the factors determined by PCA is 61.13% for the first factor (PC1), 21.52% for the second factor (PC2), 10.97% for the third

Table 2: Example of standardisation for the data of table 1 applying eqn (2).

Cd	Cr	Cu	Pb	Zn
0.642532	−0.68404	−0.96769	−1.01857	−0.94764
0.94252	−0.76207	−0.64273	−0.91538	−1.37306
0.649509	0.564402	−0.84866	0.278595	−0.98666
0.25185	0.486374	−0.31305	1.944269	−0.81422
1.842484	−1.04817	−0.60684	−0.85642	0.271995
−1.08763	−0.21588	−0.27338	0.838733	0.113405
−0.80857	−0.39794	0.215945	−0.56161	0.14613
−0.85043	1.994913	2.065549	−0.03095	1.458904
−0.85043	1.058578	1.439253	1.015619	1.256261
−0.73183	−0.99616	−0.06839	−0.69428	0.87489

Table 3: The **S** matrix reporting the scores determined by PCA.

PC1	PC2	PC3	PC4	PC5
1.8569	0.2771	0.0789	0.4575	−0.1113
2.0179	0.0538	0.3286	0.5201	0.3900
0.8499	−1.1393	0.4524	0.3508	−0.4043
−0.1478	−2.1491	−0.1513	−0.3829	0.1765
1.7715	0.8137	0.8632	−1.0594	−0.0391
−0.5415	−0.4526	−1.2251	−0.0763	−0.1261
−0.1952	0.7738	−0.6082	0.4121	0.1374
−3.0727	0.5866	1.0649	0.4188	−0.0590
−2.5169	−0.1291	0.1095	−0.3746	0.1456
−0.0222	1.3650	−0.9129	−0.2662	−0.1096

Table 4: The **L** matrix reporting the loading values for each factor.

	PC1	PC2	PC3	PC4	PC5
Cd	0.4504	−0.1109	0.7495	−0.4640	0.0889
Cr	−0.4665	−0.3415	0.5325	0.3983	−0.4728
Cu	−0.5346	0.2133	0.2988	0.0583	0.7590
Pb	−0.2943	−0.7676	−0.2557	−0.4871	0.1466
Zn	−0.4551	0.4861	0.0102	−0.6209	−0.4135

factor (PC3), 5.41% for the fourth factor (PC4) and 0.96 for the fifth factor (PC5). These results obtained for the data set of table 1 can be verified easily by means of any appropriate software package.

It should be noted that, though the score matrix reports five factors, the same number of the original five variables (Cd, Cu, Cr, Zn and Pb) coming from bio-monitoring data, only the first three factors explaining the 93.62% of the total variance are significant really whereas the remaining two last factors are noise only. This finding shows how only three new variables obtained by PCA include almost all the original information (i.e. variance) included in the biomonitoring data set with a reduction of two variables with respect to the original five variables of table 1.

This reduction gives an evident advantage: we can examine the samples of table 1 to verify similarities and differences by means of a three-dimensional plot whereas this is impossible by means of the initial five variables of table 1. This plot reported in fig. 1, shows a fundamental result. There is a difference between the two mollusc types because all the samples are placed in different positions within the plot with a marked separation between accumulation properties of the two molluscs.

This means that the two types of molluscs have a different ability in the accumulation of trace metals. The same result is not easy retrievable by means of a visual examination of table 1 data or by a simple comparison of a univariate test such as a *t*-test for a single variable. PCA shows a marked difference between the two types of molluscs based on the whole variable set and this is an advantage arising from a multivariate test such as PCA with respect to a univariate test.

However, PCA gives additional information useful for differentiating better the accumulation properties of the two molluscs. Looking at the loading values of table 4, we can see that in the first factor all the metals with the exception of Pb have a loading value ranging between 0.45 and 0.53 as absolute value while Pb has a higher loading value than all the other metals in the second factor. Taking also into account that loadings are the statistical weights of the original variables, this result shows the different contribution to the distribution of Pb in the molluscs. In fact, a re-examination of the score plot allows to see how the positions of *Patella caerulea* samples depend on a higher value of the second factor with respect to the first factor whereas the position of *Monodonta turbinata* samples depend on higher values of the first factor with respect to the second one. As previously

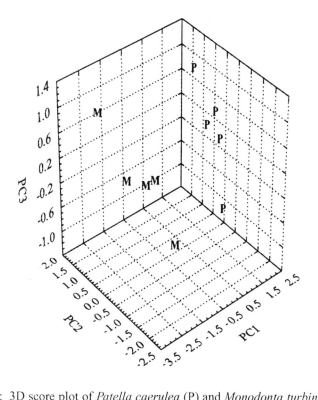

Figure 1: 3D score plot of *Patella caerulea* (P) and *Monodonta turbinata* (M).

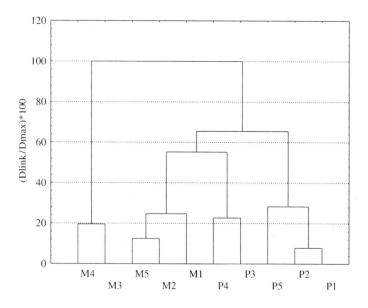

Figure 2: Cluster analysis (Ward's method, Euclidean distance) of data from table 1.

observed, Pb is the metal which determines the difference between samples in the two factors and as a consequence, *Patella caerulea* has a different accumulation of Pb with respect to *Monodonta turbinata*.

So, the case of study discussed here is an evidence of the specific insight into the structure of biomonitoring data attainable by PCA. In fact, the differences between the two molluscs can be also retrieved by Cluster Analysis (fig. 2) often applied for studying similarities and differences between data. In any case, it is evident that the dendrogram of fig. 2 shows the difference between the accumulation of trace metals between the two molluscs only but it does not show the elements which determines this difference as shown by PCA instead.

3   Supervised pattern recognition methods

3.1   Application of Hotelling's T^2 test

The application of PCA has shown the different accumulation properties of the two molluscs but we can ask ourselves whether the interpretation of PCA results is correct and the separation between the accumulation properties of the two considered species is really significant. This question concerns to verify if the two data set of *Patella caerulea* and *Monodonta turbinata* are statistically different. In an univariate case of study, the presence or absence of differences between mean values of two data sets can be verified by means of a *t*-student test whereas, for a multivariate case of study such as the data from a biomonitoring of trace metals the Hotelling's T^2 test can be applied [3]. A routine for Hotelling's T^2 test is not often present in statistical software packages however, it can be easily performed by advanced programming languages such as MATLAB (http://www.mathworks.com) or Octave (http://www.octave.org) or by means of a common spreadsheet such as Excel for Windows which includes all the mathematical functions necessary for Hotelling's T^2 test. For this reason, we describe its algorithm briefly as reported by Manley [3].

Given the two **A** and **B** matrices which represent the *Patella caerulea* and *Monodonta turbinata* data, respectively (table 1), the first step is the determination of the covariance matrix of **A** and **B** according to

$$C_1 = \frac{1}{(n_1 - 1)} \sum_{n=1}^{n_1} (\mathbf{A} - \mathbf{A}_m)'(\mathbf{A} - \mathbf{A}_m) \tag{3a}$$

$$C_2 = \frac{1}{(n_2 - 1)} \sum_{n=1}^{n_2} (\mathbf{B} - \mathbf{B}_m)'(\mathbf{B} - \mathbf{B}_m) \tag{3b}$$

where $(\mathbf{A} - \mathbf{A}_m)$ is the vectors of difference between the **A** matrix and the mean value and $(\mathbf{A} - \mathbf{A}_m)'$ is the transpose of the same vector and analogously are the terms $(\mathbf{B} - \mathbf{B}_m)$ and $(\mathbf{B} - \mathbf{B}_m)'$.

Then the pooled estimate of the global data set (i.e. table 1) is determined according to

$$C = \frac{[(n_1 - 1)C_1 + (n_2 - 1)C_2]}{(n_1 + n_2 - 2)} \tag{4}$$

where n_1 and n_2 are again the number of samples in the **A** and **B** matrices, respectively.

The Hotelling's T^2 test is defined as

$$T^2 = \frac{n_1 n_2 (A_m - B_m)C^{-1}(A_m - B_m)'}{(n_1 + n_2)} \tag{5}$$

where $A_m - B_m$ are the vectors with the mean values of variables in the **A** and **B** matrices, C^{-1} is the inverse matrix of **C** matrix previously determined by eqn (4) and the term $(A_m - B_m)'$ is the transpose of the value vector $(A_m - B_m)$.

At last, the final computation is the determination of the F-Snedekor value according to

$$F = \frac{(n_1 + n_2 - p - 1)T^2}{[(n_1 + n_2 - 2)p]} \tag{6}$$

where p is the number of variables in the data set. The experimental F value has to be compared with the two-tailed critical tabulated F value with p and $(n_1 + n_2 - p - 1)$ degrees of freedom at the selected level of significance. If the experimental F value is greater than the corresponding critical value, the hypothesis of significant differences between the two data set is verified; vice versa, if the experimental F value is lower than the corresponding critical value, the hypothesis of significant differences is rejected.

As previously reported, mean covariance, inversion and transpose of matrix are mathematical functions present in the spreadsheet Excel, so that the execution of the Hotelling's T^2 test becomes a simple numerical exercise, reported in Appendix A.

For our case of study, it is easy to verify that the experimental F is 35.88 (with 5 and 4 degrees of freedom), larger than the tabulated critical 6.26 value at the 95% of statistical significance.

This result confirms the well-separated accumulation properties of the two types of molluscs. All the computational steps are reported in Appendix A.

3.2   Discriminant analysis by Mahalanobis distance (DAMD)

The data reported in table 1 represent a set of samples taken reasonable within a single sampling step. Sometimes another set of samples with lower size or taken in a different sampling site could be available. So we could wish to verify if these new samples have the same characteristics of samples in table 1 already analysed by means of PCA.

If sample dimension of this new set is comparable with the data dimension in table 1, it will be necessary to repeat PCA including the new set, looking at the positions of these new samples in the score plot. If sample dimension of this set is lower than the original data set we can perform a discriminant analysis. Discriminant analysis is a family of multivariate techniques which are well described in some excellent textbook of multivariate analysis [3, 4] so the examination of all these techniques is out of the aim of this paper. Vice versa, our aim is to describe a simple approach available for a case of study such as the data of table 1. This approach consists of calculating the Mahalanobis distance which is the measure of the peculiar distance between an unknown sample and the centre of each group previously considered.

The Mahalanobis distance D^2 is defined according to

$$D_i^2 = (\mathbf{u} - \mathbf{x}_{im})\mathbf{C}^{-1}(\mathbf{u} - \mathbf{x}_{im})' \tag{7}$$

where $\mathbf{u}$ is the row data vector of the unknown sample, x_{im} the mean vector of the ith group (see $\mathbf{A}_m$ and $\mathbf{B}_m$ of Hotelling's T^2 test) and $\mathbf{C}^{-1}$ the inverse of the pooled sample covariance matrix of the two groups, already examined by eqn (4). At last, the term $(\mathbf{u} - \mathbf{x}_{im})'$ is the transpose of the term $(\mathbf{u} - \mathbf{x}_{im})$.

Again, as already seen for Hotelling's T^2 test, the determination of D_i^2 can be also performed by the spreadsheet Excel. When the two values of D^2 have been calculated, the unknown sample is allocated to the group where D^2 has the smallest value. Two practical examples of D^2 determination are reported in Appendix B.

3.3   Linear discriminant analysis (LDA)

Looking at fig. 1 reporting the three-dimensional plot of PCA factors, we can see (or believe to see) a separation between the two classes of samples represented by *Monodonta* and *Patella*, respectively and then we can ask ourselves whether this separation exists (i.e. it is really significant) or it is only the effect of the graphical representation used.

How can we give an answer to this question? A general answer to this question needs to verify the existence of a boundary zone between the two zones where the two typologies of samples are located and this arises from the family of statistical techniques called discriminant analysis. This family of statistical techniques has the aim to verify the existence of a significant differentiation among the typologies of examined samples, differentiation which determines a separation in the plot of data.

The term "family of statistical techniques" means that several and sometimes complex algorithms exist to identify (if it exists really) the mathematical function indicating the separation among two or more classes of samples and in this section we will examine the LDA approach consisting of the identification of the line function which describes the boundary line between two classes of samples such as the case of metals present in *Monodonta* and *Patella* molluscs. The approach of LDA was developed by Fisher [4] and it is based essentially on the use of variance–covariance matrix of the original data and the related score matrix determined by PCA.

We reconsider the eqns (3) and (4) concerning the determination of the covariance matrices of the groups of samples named **A** and **B**, respectively (eqn (3)), and the covariance matrix of the two classes (eqn (4)).

Then we can determine the LDA function X according to

$$X = (\mathbf{A}_m - \mathbf{B}_m)\mathbf{C}\mathbf{E}' \tag{8}$$

where $\mathbf{A}_m$ and $\mathbf{B}_m$ are the mean vectors of **A** and **B** data set, **C** is the pooled covariance matrix determined according to eqn (4) and $\mathbf{E}'$ the transpose of the data matrix including both **A** and **B** matrix.

If the first two factors of PCA explain a significant fraction of total variance (i.e. higher than 70%) we can plot the two first factors and then use the first factor of PCA to plot the **X** vector in the same two factor scores. This allows to verify if the **X** vector show a boundary zone (i.e. line) within the two classes of samples. The example of LDA analysis for the data of table 1 is reported in the appendix, Section C.

4   Conclusions

The presented approaches using multivariate analysis are useful tools for data interpretation of biomonitoring studies. In particular these consent to clearly observe the variables which determine differences and similarities within the studied species and in addition, they also allow the classification and discrimination among the properties of each typology of species.

At last, it should be noted that the application of the multivariate techniques reported here can be performed either by means of common software packages or by means of common spreadsheet such as Excel or high level computer languages such as MATLAB or Octave.

Appendix A: Hotelling's T^2 test for data of table 1

Given the two matrices of *Patella caerulea* (**A**) and *Monodonta turbinata* (**B**)

$$
\mathbf{A} = \begin{vmatrix}
3.25 & 0.27 & 2.78 & 0.4 & 41.78 \\
3.68 & 0.24 & 6.22 & 0.47 & 38.4 \\
3.26 & 0.75 & 4.04 & 1.28 & 41.47 \\
2.69 & 0.72 & 9.71 & 2.41 & 42.84 \\
4.97 & 0.13 & 6.6 & 0.51 & 51.47
\end{vmatrix}
$$

$$\mathbf{B} = \begin{vmatrix} 0.77 & 0.45 & 10.13 & 1.66 & 50.21 \\ 1.17 & 0.38 & 15.31 & 0.71 & 50.47 \\ 1.11 & 1.3 & 34.89 & 1.07 & 60.9 \\ 1.11 & 0.94 & 28.26 & 1.78 & 59.29 \\ 1.28 & 0.15 & 12.3 & 0.62 & 56.26 \end{vmatrix}$$

The covariance $\mathbf{C_1}$ and $\mathbf{C_2}$ of $\mathbf{A}$ and $\mathbf{B}$ matrices are

$$\mathbf{C_1} = \begin{vmatrix} 0.7362 & 0.1860 & 0.1907 & 0.4700 & 3.0894 \\ 0.1860 & 0.0845 & 0.1842 & 0.2107 & 0.5000 \\ 0.1907 & 0.1842 & 7.0745 & 1.5532 & 2.6321 \\ 0.4700 & 0.2107 & 1.5532 & 0.7366 & 0.4119 \\ 3.0894 & 0.5000 & 2.6321 & 0.4119 & 24.1429 \end{vmatrix}$$

$$\mathbf{C_2} = \begin{vmatrix} 0.0364 & 0.0085 & 0.4462 & 0.0720 & 0.4045 \\ 0.0085 & 0.2173 & 4.7924 & 0.1033 & 1.6607 \\ 0.4462 & 4.7924 & 117.121 & 1.2769 & 45.4321 \\ 0.0720 & 0.1033 & 1.2769 & 0.2841 & 0.2687 \\ 0.4045 & 1.6607 & 45.4321 & 0.2687 & 24.339 \end{vmatrix}$$

According to eqn (3) the pooled estimate $\mathbf{C}$ is

$$\mathbf{C} = \begin{vmatrix} 0.3863 & 0.0972 & 0.1278 & 0.2710 & 1.7469 \\ 0.0972 & 0.1509 & 2.4883 & 0.1570 & 0.5804 \\ 0.1278 & 2.4883 & 62.0978 & 1.4150 & 24.034 \\ 0.2710 & 0.1570 & 1.4150 & 0.5104 & 0.0716 \\ 1.7469 & 0.5804 & 24.032 & 0.0716 & 24.241 \end{vmatrix}$$

The inverse of $\mathbf{C}$ used in the eqn (4) is

$$\mathbf{C^{-1}} = \begin{vmatrix} 11.82 & 18.964 & 0.1232 & 3.5188 & 0.9503 \\ 9.6442 & 46.865 & 1.7671 & 4.4053 & 0.0781 \\ 0.1232 & 1.7671 & 0.1020 & 0.1885 & 0.0493 \\ 3.5188 & 4.4053 & 0.1885 & 4.6152 & 0.3213 \\ 0.9503 & 0.0781 & 0.0493 & 0.3213 & 0.1596 \end{vmatrix}$$

The mean vectors $\mathbf{A}_m$ and $\mathbf{B}_m$ of $\mathbf{A}$ and $\mathbf{B}$ are

$$\mathbf{A}_m = 3.5700 \quad 0.4220 \quad 5.8700 \quad 1.0140 \quad 43.1920$$
$$\mathbf{B}_m = 1.0880 \quad 0.6440 \quad 20.1780 \quad 1.1680 \quad 55.4260$$

According to eqn (3) T^2 value is 358.84 and according to eqn (4) the experimental F value is 35.8.

Appendix B: Discriminant analysis by Mahalanobis distance for data of table 1

In the first example, we have a *Patella caerulea* sample with a metal content of Cd, Cr, Cu, Pb and Zn concentration 3.60, 0.50, 3.80, 0.70 and 40.50 µg g^{-1}, respectively. We will consider this sample as an unknown because we wish to verify if it has the characteristics of *Patella caerulea* samples of table 1

The $\mathbf{C}^{-1}$ matrix has been calculated for Hotelling's T^2 test already.

The term $(\mathbf{u} - \mathbf{x}_{im})$ of eqn (5) for the first sample has values (0.0300 0.0780 –2.0700 –0.3140 –2.6920). By means of eqn (5), the Mahalanobis distance between the unknown and *Patella caerulea* group is 2.46 and the Mahalanobis distance between the unknown and *Monodonta turbinata* group is 170.63. So, the unknown sample confirms to have the characteristics of *Patella caerulea* samples.

In the second example, we have a *Monodonta turbinata* sample with a metal content of Cd, Cr, Cu, Pb and Zn concentration 1.10, 0.90, 28.00, 1.50 and 50.00 µg g^{-1}, respectively.

The term $(\mathbf{u} - \mathbf{x}_{im})$ of eqn (5) for the first sample has values (0.0120 0.2560 7.8220 0.3320 –5.4260). The Mahalanobis distance between this sample and *Monodonta turbinata* group is 13.42 whereas Mahalanobis distance between this sample and *Patella caerulea* group is 102.58; so this sample confirms to have the characteristics of a *Monodonta turbinata* sample.

Appendix C: Example of LDA for data of table 1

If we apply eqn (8) taking into account all the herein terms previously determined by the other equation we can have the following values for the $\mathbf{X}$ vector representing the LDA function:

$$\mathbf{X} = 53.9817 \quad 79.9186 \quad 82.8973 \quad 61.0524 \quad 80.9955 \quad -60.8459 \quad -62.4344$$
$$-76.4587 \quad -72.7686 \quad -86.3$$

Plotting the $\mathbf{X}$ function with respect to the first factor of PCA and including this plot in the same plot of the first two factors we can obtain a final plot reported in fig. 3, showing the presence of a significant boundary line between *Monodonta turbinata* and *Patella caerulea* samples.

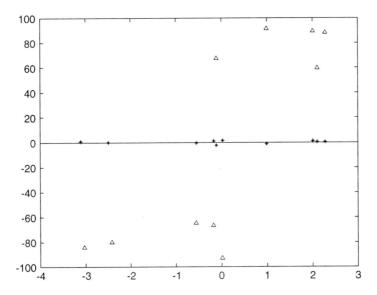

Figure 3: LDA plot for the data of table 1. "Δ" shows the scores data of the first two factors and "*" shows line of discriminant analysis.

References

[1] Conti M.E., Iacobucci M., Cecchetti G. (2005) A statistical approach applied to trace metal data from biomonitoring studies. International Journal of Environment and Pollution, 23(1), 29–41.

[2] Meglen R.R. (1992) Examining large databases: a chemometric approach using Principal Component Analysis. Marine Chemistry, 39, 217–237.

[3] Manley B.F.J. (1986) Multivariate Statistical Methods: A Primer. Chapman & Hall, London, UK.

[4] Brereton R.G. (1990) Chemometrics, Application of Mathematics and Statistics to Laboratory Systems, Ellis Hollis Horwood Series in Chemical Computation, Chichester, UK.

[5] Massart D., Kaufmann L. (1983) The Interpretation of Analytical Chemical Data by the use of Cluster Analysis, Krieger Publishing Company, Malabar, FL, USA.

[6] Bisani M., Clementi S., Wold, S. (1982) Elementi di Chemiometria. La Chimica e l'Industria, 64, 655–665.

[7] Geladi P. (2002) Calculating Principal Component Loadings and Scores in MATLAB, ISBN 91-7191-083-2, Umeå, Sweden.

[8] Conti M.E., Iacobucci M., Mecozzi M., Cecchetti G. (2006) Trace metals in soft tissues of two marine gastropod molluscs: *Monodonta turbinata* B. and *Patella caerulea* L. collected in a marine reference ecosystem, in: Brebbia, C.A. (Ed.), Environmental Problems in Coastal Regions VI, Including Oil Spill Studies. WIT Press, UK, pp. 3–11.

[9] Conti M.E., Iacobucci M., Cucina D., Mecozzi M. (2007) Multivariate statistical methods applied to biomonitoring studies. International Journal of Environment and Pollution, 29(1–3), 333–343.

[10] Meloun M., Militký J., Hill M., Brereton, R.G. (2002) Crucial problems in regression modelling and their solutions. Analyst, 127, 433–450.

Index

...for scientists by scientists

Environmental Toxicology II

Edited by: A. KUNGOLOS, University of Thessaly, Greece, C.A. BREBBIA, Wessex Institute of Technology, UK and M. ZAMORANO, University of Granada, Spain

The science of environmental toxicology is one of the most interdisciplinary ones. Biologists, microbiologists, chemists, engineers, environmentalists, ecologists, and other scientists have worked hand in hand developing this new discipline. The issue of the assessment of environmental effects of chemicals is complicated; it depends on the organism tested and involves not only toxicity testing of single chemicals, but also interactive effects (including synergistic ones) and genotoxicity, mutagenicity and immunotoxicity testing.

The issue of hazardous waste management is also closely related to environmental toxicology, and there is a growing need for techniques and practices to minimize the environmental effects of chemicals and for the implementation of the corresponding principles in the planning of environmental policy and decision-making. With many new chemicals entering the market every year, it has become necessary to assess their effects on the ecosystem, and to find ways to minimize their impact on the environment.

The Second International Conference on Environmental Toxicology brought together a wide range of people working within the many disciplines associated with environmental toxicology and hazards waste management, including: biologists; environmental engineers; chemists; environmental scientists; microbiologists; medical doctors; and all academics, professionals, policy makers and practitioners.

This book contains the papers presented at the Conference and covers subject areas such as: Risk Assessment; Effluent Toxicity; Pharmaceuticals in the Environment; Genotoxicity; Bioaccumulation of Chemicals; Monitoring, Assessment and Remediation of Hazardous Waste; New Trends in Environmental Toxicology; Bio Tests; Biodegradation and Bioremediation; Ecotoxicity of Emerging Chemicals; Soil Ecotoxicity; Protyomics and Genomics; Biological Effects Monitoring; Endocrine Distribution in Wildlife; Laboratory Techniques and Field Validation; Online Toxicity Monitoring; Ecosystem and Human Health Assessment; Exposure Pathways; Integrated Approach to Risk Assessment; Bio-terrorism Safety and Security; Habitat Destruction.

WIT Transactions on Ecology and the Environment, Vol 110
ISBN: 978-1-84564-114-6 2008 apx 300pp apx £95.00/US$190.00/€142.50

 **WIT**_PRESS_   *...for scientists by scientists*

Environmental Health Risk IV

Edited by: **C.A. BREBBIA**, *Wessex Institute of Technology, UK*

Health problems related to the environment have become a major source of concern all over the world. The health of the population depends upon good quality air, water, soil, food and many other factors. The aim of society is to establish measures that can eliminate or considerably reduce factors hazardous to the human environment to minimize the associated health risks. The ability to achieve these objectives is greatly dependant on the development of suitable experimental, modelling and interpretive techniques, which allow a balanced assessment of the risk involved as well as suggest ways in which the situation can be improved.

The interaction between environmental risk and health is often complex and can involve a variety of social, occupational and lifestyle factors. This emphasizes the importance of considering an interdisciplinary approach. Containing papers presented at the Fourth International Conference on The Impact of environmental Factors on Health, the book discusses topics that will be of interest to a wide readership including health specialists in government and industry as well as researchers involved within the broad area of environmental health risk. Featured topics include: Risk Analysis; Air Pollution; Water Quality Issues; Electromagnetic Fields; Food Contamination; Occupational Health; Remediation; Social and Economic Issues; Housing and Health; Radiation Fields; Education and Training; Accident and Man-made Risks.

WIT Transactions on Biomedicine and Health, Vol 11
ISBN: 978-1-84564-083-5 2007 304pp £95.00/US$185.00/€142.50

Environment and Health

Protecting our Common Future

K. DUNCAN, *University of Toronto, Canada*

Environmental degradation and illness and disease prevent millions of people in many countries from surviving and achieving their potential. This book thus serves as a comprehensive guide to key environmental and health issues confronting the planet, enumerates approaches and techniques to address these issues, and provides real-world examples of good corporate citizenship.

Although there are a growing number of books on corporate social responsibility (and related concepts, such as business ethics, stakeholder theory and sustainable development), there is no volume comparable to this one. Specifically the book addresses a range of issues that will be invaluable to many specialists such as organization leaders who want to improve the environment and health of their colleagues at home and globally, practitioners in corporate social responsibility, managers involved in environmental, health and safety issues and finally business students who work to enhance the well-being of their colleagues and the health of the planet and humanity.

ISBN: 978-1-84564-130-6 2008 apx 192pp apx £63.00/US$126.00/€94.50

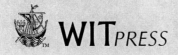

WITPRESS *...for scientists by scientists*

Disposal of Hazardous Waste in Underground Mines

Edited by: **V. POPOV**, *Wessex Institute of Technology, UK and* **R. PUSCH**, *Geodevelopment AB, Sweden*

This book contains the results of a three-year research programme by a joint team of experts from four different EU countries.

The main focus of this research was on investigating the possibility of using abandoned underground mines for the disposal of hazardous chemical waste with negligible pollution of the environment.

The research work focused on: the possibility of using abandoned underground mines for disposal of hazardous chemical waste with negligible pollution of the environment; the properties and behaviour of waste-isolating clay materials and practical ways of preparing and applying them; development of software tools to assess the stability, performance and risks associated with different repository concepts, considering the long-term safety of the biosphere; the isolating capacity of reference repositories; the different approaches for handling hazardous chemical waste. The project has demonstrated that hazardous waste can be safely disposed of in underground mines provided that adequate assessment, planning and design procedures are employed.

Information is also included on the selection of site location, design and construction of repositories, predicting degrees of contamination of groundwater in the surroundings, estimation of isolating capacity of reference repositories, cost estimation of this approach in comparison with some other approaches, and many other relevant issues.

Invaluable to researchers and engineers working in the field of hazardous (chemical) waste disposal, this title will also significantly aid experts dealing with nuclear waste.

Series: The Sustainable World, Vol 11
ISBN: 1-85312-750-7 2006 288pp £95.00/US$170.00/€142.50

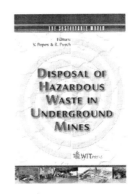

WITPRESS   *...for scientists by scientists*

Environmental Toxicology

Edited by: **A. KUNGOLOS**, *University of Thessaly, Greece*, **C.A. BREBBIA**, *Wessex Institute of Technology, UK*, **C.P. SAMARAS**, *TEI of West Macedonia, Greece and* **V. POPOV**, *Wessex Institute of Technology, UK*

This book addresses the need for the exchange of scientific information among experts on issues related to environmental toxicology, toxicity assessment and hazardous waste management. Publishing papers from the First International Conference on Environmental Toxicology, the text will be of interest to biologists, environmental engineers, chemists, environmental scientists, microbiologists, medical doctors and all academics, professionals, policy makers and practitioners involved in the wide range of disciplines associated with environmental toxicology and hazardous waste management. The text encompasses themes such as: Acute and Chronic Bioassays; Tests for Endocrine Disruptors and DNA Damage; Interactive Effects of Chemicals; Bioaccumulation of Chemicals; Assessment of Ecotoxicological Properties of Hazardous Wastes; Hazardous Waste Management Techniques; Legislation Regarding Environmental Effects of Chemicals; Hazardous Waste Reduction and Recycling Techniques; Biodegradation and Bioremediation; Monitoring of Hazardous Waste Environmental Effects; Laboratory Techniques and Field Validation; Effluent Toxicity, Microbiotests; On-line Toxicity Monitoring; Forensic Toxicology; Genotoxicity/Mutagenicity; Exposure Pathways; Risk Assessment; Biotesting and Environmental Control Strategy; Hot Spots and Accidental Spills.

WIT Transactions on Biomedicine and Health, Vol 10
ISBN: 1-84564-045-4   2006   384pp   £125.00/US$225.00/€187.50

WITPress
Ashurst Lodge, Ashurst,
Southampton,
SO40 7AA, UK.
Tel: 44 (0) 238 029 3223
Fax: 44 (0) 238 029 2853
E-Mail: witpress@witpress.com

 **WIT**PRESS *...for scientists by scientists*

Environmental Exposure and Health

Edited by: **M.M. ARAL**, *Georgia Institute of Technology, USA,* **C.A. BREBBIA**, *Wessex Institute of Technology, UK,* **M.L. MASLIA**, *ATSDR/CDC, USA and* **T. SINKS**, *NCEH, USA*

Current environmental management policies aim to achieve sustainability while improving the health, safety and prosperity of the population. This is an interdisciplinary activity that requires close cooperation between different sciences.

Featuring contributions from health specialists, social and physical scientists and engineers this volume evaluates current issues in exposure and epidemiology and highlights future directions and needs. Originally presented at the First International Conference on Environmental Exposure and Health, the papers included cover areas such as: METHODOLOGICAL TOPICS – Methods Of Linking Epidemiology, Exposure and Health Risk; Multipathway Exposure Analysis and Epidemiology; Statistical and Numerical Methods. SITE RELATED TOPICS – Work Place and Industrial Exposure; Soil, Dust and Particulate Exposure; Water Distribution Systems, Exposure and Epidemiology; Air Pollution Exposure and Epidemiology. DATA COLLECTION TOPICS – Use of Remote Sensing and GIS; Data Mining and Applications in Epidemiology. SPECIAL TOPICS – Exposure Specific to the Developing World; Epidemiology of Mixed Chemical and Microbial Exposure; Effects of Rapid Transportation in Epidemiology; Interaction of Social and Environmental Issues and Health Risk.

WIT Transactions on Ecology and the Environment, Vol 85
ISBN: 1-84564-029-2 2005 528pp £185.00/US$325.00/€277.50

WIT eLibrary

Home of the Transactions of the Wessex Institute, the WIT electronic-library provides the international scientific community with immediate and permanent access to individual papers presented at WIT conferences. Visitors to the WIT eLibrary can freely browse and search abstracts of all papers in the collection before progressing to download their full text.

Visit the WIT eLibrary at
http://library.witpress.com